All The World Is A Stage

Is A Stage

& Every Actor Meets The Critic

YOUR LIFE! NOW PLAYING STARRING: YOU

William E. Adams

All the World
is a Stage. . .

& Every Actor Meets The Critic

William E. Adams

Illustrations by Tim Tarcha

Old Drum Publishing

Box 401

Portersville, PA 16051

Order forms in back of book

First printing, April, 1997

Library of Congress Catalogue Card #:

ISBN 0-9639623-2-9

100% USA

Acknowledgments:

My wife, Eileen, has been a great help in this endeavor, as have my parents and my children. Special thanks to my Grandfather Charles and to my Aunt Louise.

The monks of St. Bede's School, particularly Fr. William, Fr. Gordian, Fr. Theodore, Fr. Owen, Fr. Albert, Fr. Damien, and Mr. T. Bower Campbell introduced me to a world of thought so beautifully complex that I have only been able to scratch its surface.

Almost twenty years after I left St. Bede's, Fr. Ernest Strzelnski received me into the Church at St. Christopher's. Today, the wonderful Pastor of St. Christopher's, Fr. Matthew Tosello, continues to offer guidance that is both sound and indispensable.

My friend, Charlie Goetz, helped me to make a smooth transition from business interests to hacking out this crude trail through thick underbrush. Thank you, Charlie.

"Eliminating mental clutter and confusion raises our IQ to its highest natural capacity.

"To be as smart as we can be, we only need to find the simplest way that uncontradictably describes the world."

Author's notes

"Watson, it doesn't matter to me, or to my work, if the earth goes around the sun or if the sun goes around the earth."

Sherlock Holmes

"Everything unnecessary should be cut away."

St. William of Ockham

Warning-Disclaimer:

Throughout the long succession of miracles that make up the Jewish and Christian Faiths of our Fathers, professionals have been provided to give guidance.

Salvation is not a do-it-yourself project. Simply reading this, or any book, is no guarantee. Professional help from ordained, traditional clergy can be a great help.

Seek no guidance from anyone who does not believe that God is capable of miracles. If, at Judgment, we are denied admittance to Heaven, it will be, to some degree, because we did not bother to select a good Guide.

"Row, row, row your boat
gently down the stream.
Merrily, merrily, merrily, merrily,
Life is but a dream."

American song

Preview

Your Computer. A Step to God?

You know how screen savers, video games, and computer simulations look more real every year.

God is so much smarter that He can program actual shapes that make things we think are real.

Each screen-saver, video game, and virtual reality scenario is a reminder that all Creation easily could have been programmed out of three-dimensional fractals. In a week. By a God far smarter than we.

Getting Over Conventional Reality

Plato taught that most people go through life and only see shadows. Only a few want to see the Light that *makes* the shadows that form *Conventional Reality.*

Aquinas understood that the true *substance* of a thing is different from its apparent *form.*

People who study Genetics know there's a

difference between *genotype*, what a thing really is, and *phenotype*, what it appears to be.

Simply saying "That's Conventional Reality" when we hear things on the "news" begins our ascent.

Conventional Reality is mostly comprised of Press Releases. We can get over it.

Simplify. Simplify. Simplify.

A brain cluttered with Conventional Reality can't work any better than a car with extra wheels. IQ goes up as confusion goes down. The easiest way to describe life and where we live it? Shakespeare's very possible conclusion: "All the world's a stage."

Then

Consider the reasonable and uncontradictable conclusion that Shakespeare is literally right, that the world actually is a Stage. It's an easier conclusion to reach once we consider that an Invisible Producer could have easily made a Stage and Actors who could make Actors out of 3-D Fractals, actual shapes programmed in what used to be formless void.

Just Imagine!

All the world's a stage. It's
just a stage!

Could it be?

The Curtain Opens

Only Two Things Can be Seen on Stage: Props & Actors

Props

Most of us believe that the Universe is big, old, and complicated. That's Conventional Reality, and we're used to it. It's hard to see that we may be simply Actors on a Stage made out of Props.

Most Actors are taught to reject that possibility with knee-jerk speed. Today, it is popular among the Actors to believe that we live on a large, rotating ball. It is not fashionable to believe that the large, rotating ball is a Stage. If we Consider that it might be, two recurring things are noticeable:

1. During every recent performance, tinier Props have been found. When a tinier Prop is found, like an electron, the Actor who found it

gets to stand in a prominent place and announce dramatically to the other Actors: "We are making progress!"

2. Actors frequently find Props that appear

to be older, or larger, or farther away. Whenever a larger, older, or more distant Prop is found, like a quasar, the Actor who found it gets to stand in a prominent place and announce dramatically to the other actors: "We are making progress!"

What are Props Made of?

No one knows. The simplest things that Props could be made of are 3-D Fractals. No one knows what they are, either. An easy way to picture 3-D Fractals is to visualize a Screen Saver on a computer.

Men can program moving shapes in two dimensions on Screen Savers. God is so smart that He can program actual shapes in actual space. When He says "Sand.", whole beaches of it can spread automatically around the world. When He says "Sandstone", layers of it appear wherever He wants them to be.

Such Props are layered and veined so realistically that many Actors reasonably choose to believe that they appeared by accident, long ago. Or, Actors may also reasonably choose to believe that the layers of rock, pools of oil, fossils, and other old-looking features were simply programmed in place a few thousand years ago as set decorations.

Why?

Asking why an Invisible Producer would have made The Stage look so old and complicated leads us to a Possible Conclusion: *Creation is simply the simplest way that each new generation of Creatures with Free Will can be given lots of opportunities to choose good or evil.*

The Two Kinds of Props

The Basic Props are earth, air, fire, and water. Today known as solids, liquids, gasses, and energy, they seem to exist without Actors. The other kind of Props are Doodads. Doodads, like airplanes, mink coats, and ideas, exist because Actors relocated or rearranged Basic Props.

It can be hard to tell Basic Props from Doodads. It's difficult to tell a cubic zirconium from a diamond without a powerful microscope.

Doodads are made by "Makers".

Doodads

Some Actors have a knack for putting Props together in new ways. They love to recite lines containing the word "Eureka!", which means "I've got a New Doodad!"

Actors who think up New Doodads make other actors nervous, especially those who make a living with Old Doodads.

All economic struggles are the result of conflicts between the few Actors who say "Eureka!" and the vastly larger number of those who don't.

New Doodads become Stage Props for current performances. Doodads that last a long time become Props for future performances. Each Current Performance calls old Doodads "antiques". Doodads outside the mind can be as big as Great Walls, as simple as Shaker chairs, or as bright and shiny as gold doubloons.

Doodads inside the Mind are called Ideas. They can be as complicated as whole philosophies or as simple as a legal argument.

Doodads Inside the Mind

Among the teeny-weeniest of Basic Props are the little things inside the mind that can be arranged to make Thoughts. Thoughts can be piled up to form Ideas, Opinions, Beliefs, and Attitudes. All are Doodads In the Mind.

A Deal is an advanced type of such a Prop. Some Actors don't make anything but Deals. Example: Actors in Real Estate don't actually make land. They arrange exchanges of Titles to Props for Symbols for Props. Such exchanges are "Deals".

Deals involve Time. A maker of a deal must be smart enough to think ahead in time to analyze future events that can affect the Deal. In deals, stability is crucial. The Biggest Deals are called Treaties.

Money: $ymbols for Props

Makers have been cranking out Doodads since the two Actors in the First Run made Costumes out of leaves. Owning too many Doodads paralyzes the Actors.

On some parts of The Stage, dead Actors had to have their Doodads buried with them, just to get excess Doodads off The Stage. Some Doodads are so big that even Movers can't get them from place to place.

Actors realized that exchanging Props and Doodads had become too burdensome. They decided to trade $ymbols for Props, rather than exchanging actual Props.

"I'll pluck the feathers off this dead chicken, clean it, wrap it, and give it to you if you'll give me a $ymbol for Props that we'll call a dollar." was much easier for everyone than "Og, I'll give you six clams for your rooster if you get it plucked and packed."

Actors on the Stage call $ymbols for Props "money" or "cash", among other things.

How Props are Exchanged

Some Props are exchanged by Free Trade. Props can also be Taken by Force. Forced Exchange is called "Taxation", "Confiscation", or "Theft", depending on which Actors are Giving Opinions.

About the Actors
Actors Only Have Two
Kinds of Characteristics

Actors' Characteristics are either Given or Chosen. Given Characteristics, like size or intelligence, are gifts. Good manners require that we say "Thank you" for gifts. Instead, many Actors foolishly congratulate themselves for having the Gifts they have been given. "I am so smart, quick, healthy, and live in a great Section of the Stage!"

7

EVERY ACTOR IS A STAR
In a Play that each Actor Calls
"My Life"

"My Life" is a play in which each Actor (there are several billion) is The Star. In "My Life", all Actors are forced to use Characteristics they've been Given to Choose and Use the Characteristics they decide to have.

At the end of the Play, Actors who choose characteristics like honesty, diligence, kindness, humility, chastity, and generosity are rewarded differently than Actors who choose deceit, sloth, anger, arrogance, lustfulness, and greed.

The Actors on every Section of The Stage are Organized in a Very Simple Way:

A. EVERY ACTOR IS IN A TROUPE.

B. EVERY TROUPE IS IN A GUILD.

There are only three Guilds. Every Actor on The Stage is either in:

1. The Workers' Guild

2. The Lubricators' Guild

3. The Guild of Government

THE WORKERS' GUILD
Working With Props

History of the Workers. Once, Actors didn't need to work. They became Workers after they got themselves thrown off the Original Set for not following Directions. Now, most Actors have to work. Workers are in six main Troupes:

Growers/Gatherers. G/Gs work with Props that grow on the stage. The Props with which they work are, or were, alive. Their Props are usually eaten, worn, driven around on, burned, or sawed.

Miners/Separators. They pump, dig, drill, and burrow to bring mineral Props to the Walking Around Level of The Stage.

They also Separate mineral Props according to desirability and put them into piles, tanks, and chunks.

Movers/Storers. One kind of Mover/Storer handles Props. They deal with freight trains, cargo ships, and warehouses.

The other kind of Mover/Storer handles Actors. Airline pilots and cab drivers move Actors to different parts of The Stage and carry costumes for short runs. Hotels and motels store Actors for short times. Apartments and houses store Actors for longer periods.

Makers. Makers turn Props into Doodads. Some form thought particles into ideas that can be made into products.

Fixers. One kind of Fixer takes care of Props. The others fix and maintain Actors.

Buyers/Sellers. They buy and sell Basic Props, Doodads, and Performances. The more useless the Doodad, the better the Performance needed to sell it.

Workers have Supporting Casts of specialized, interchangeable types of Workers. The Supporting Casts include: **Counters, Checkers, & Managers.**

THE LUBRICATORS
Thinking, Talking, and Billing

History of The Lubricators. As more Actors invented more ways to move more Props around The Stage with more speed, there were more collisions. At first, colliding Actors would try to kill each other. That was not efficient. "Let's see if we can work things out." said the first Lubricator.

When they give their soothing Performances on The Stage, Lubricators play Roles as clergymen, lawyers, agents, judges, lobbyists, counselors, adjusters, consolidators, consultants, and more. They are called less complimentary things by people who are not well oiled. There are four Troupes of Lubricators.

Public Lubricators. When Actors got tired of running into each other, Directors appeared to give Directions. Actors call these Directions "laws".

Laws come from three places:
1. Words said to come from God.
2. Past practice.
3. Current need.

Laws are applied to Actors by an elaborately costumed Lubricator called "Judge". "Judges" perform in elaborate sets called "Courts".

Other Lubricators of Public Problems work to influence, change, or carry out the Judges' Directions. They include bailiffs, tipstaffs, clerks, lobbyists, attorneys, barristers, judges of lower courts, magistrates, sheriffs, deputies, and parole officers.

Private Lubricators. These are called "attorneys", "lawyers", "counselors", "agents", "personnel officers", and "arbitrators". As Lubricators of Public Problems deal with ever larger and more complicated collisions, many Actors seek to solve their problems by keeping them out of the hands of the dreaded Public Officials.

Lubricators of Private Problems do not have significant Power given to them by Public Officials.

Holy Oilers. These are known on Stage as "Priests", "Ministers", or "Rabbis". Holy Oilers seek to reduce friction between Actors who ask for their help. They also attempt to smooth frictions between Actors and the Invisible Producer.

The best Holy Oilers love the Invisible Producer. They try to become more familiar with His Directions.

Truly Holy Oilers believe the Invisible Producer outranks Public Officials. For that reason, large numbers of Holy Oilers are often removed from The Stage by Public Officials.

De-Lubricators. Special Interests use them to cause friction and spread confusion among competing interests.

In Perpetual Battle against
All Other Actors
is the powerful, complicated

GOG

GUILD OF GOVERNMENT

History of the GOG: In Earlier Performances, Actors lived in small units called
"families". Each family fed and clothed its
members with Props that they found, grew,
and made.

As Families combined into Clans and Tribes,
more Doodads were invented, and efficency
increased. The invention of the spinning
wheel let a few Actors produce cloth for many.
One potter with a wheel could make enough
containers for a village. When two or more
wheels were attached to a platform, a dozen
Actors could move enough Props around for
all the Actors in town. Huge bows and slings
on wheels allowed a few Actors to kill cities
full of Actors.

When different kinds of wheels were hitched up to Yokes, even fewer Actors were needed to make, grow, find, and move Props for other Actors. First came wheeling. Then, dealing.

Universal Constant: When every Actor had to work full time just to get the basics, there was little left to grab. *The Guild of Government increases in proportion to the amount of production overcapacity.* The more efficiently goods and services are produced, the more there is to grab, and the bigger the GOG becomes.

Those left without Roles by new Doodads either lose status, die, invent better Doodads, or live by getting Props from the GOG.

Troupes of the Guild of Government are Supporting Casts for Public Officials. Some of the GOG's Troupes are Good. Some are Pawns. Some are Bad.

The Good GOGsters

Civil Servants. Civil Servants tend to be civil and to serve. Bad GOGsters despise them. Civil Servants are honest, hard-working Actors who try to put on an honest days' performance for reasonable access to Props.

Civil Servants and Workers have much in common. Low-level Workers often marry Civil Servants to get better benefits. Hard-working Civil Servants include mailmen, patent office workers, customs' agents, soldiers, sailors, and others who work in Useful Jobs that have been mandated by The Constitution.

Small-Caliber Protectors. These Civil Servants usually wear blue or brown costumes and travel singly or in pairs. They often decorate their costumes with a highly reflective chest-badge. Each carries a small Death Doodad, usually of .45 caliber or less. They are commonly called "policemen", "deputy", "sheriff", "officer", "Agent", or "trooper."

Large-Caliber Protectors. These heavily armed Civil Servants operate in larger, synchronized groups. They protect against Maniacally Acquisitive Guilds of Governments (MAGOG) from other sections of The Stage. Large-Caliber Protectors have huge Death Doodads and more varied means of rapidly moving them about The Stage. They are called "bombradiers", "artillerymen", "submariners", and dozens of other names that often describe the Death Doodads with which they are specially skilled.

Protectors. They usually help Actors near them on The Stage. Many times, they will risk their lives to protect other Actors.

Protectors are often forced to take orders from bad Public Officials. Sometimes, Protectors are turned into Bad GOGsters. When Protectors are totally perverted by Public Officials Gone Bad, they turn into MAGOG, Maniacally Acquisitive Guilds of Government. As such, they will remove whomever they're told to remove from The Stage, as at Waco.

Pawns of The GOG

Tubers.　Tuberfellas and Tuberellas spend most of their waking hours staring into the Vacuum Tubes from which they get their names. Tubers live on Freebies.

Tubers get Freebies by Dribble Down, a complicated process, called "Welfare", among other things.　Tubers are needed to make "jobs" for Welfare and Social "Workers", as they like to call themselves.

Distractors.　Workers must be given things to distract them from higher taxes, increased regulations, and their descent into slavery.

Ideally, Distractors provide things that are complex enough to get actors to distract each other by talking about them.

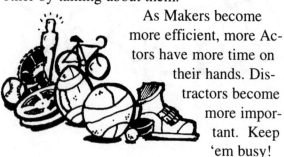 As Makers become more efficient, more Actors have more time on their hands. Distractors become more important. Keep 'em busy!

Many distractions turn into professions. Tall Distractors play basketball. Burly Distractors play football. Other Professional Distractors play baseball, golf, and tennis. Some are in moving pictures. Many put words in approved books, magazines, and newspapers.

Public Officials give Important Distractors special privileges, like free stadia, and exemption from anti-trust laws. Often, they're given tax money. In turn, the "owners" of these organizations provide Distractions.

L'il Grabbers (criminals). Tuberellas are given more Freebies each time they give birth to a child in a fatherless home.

Such children are viciously dumbed-down by Meddlers who force Young Tubers to live in poverty, without fathers, unable to learn right from wrong, dependent on the dreaded Public Housing.

Young Tuberfellas and Tuberellas are encouraged to fornicate outside of marriage to produce lots of L'il Grabbers. As soon as a Tuberella produces her first fatherless child, Meddlers of the GOG reward her with her own apartment, free food, medicine, and utilities.

Many such children are subtly taught that

they have "a right" to snatch Props by force or by stealth from Actors in their own and other Guilds. They are not taught ways to obtain Props by Free Exchanges.

Media Frauds & Fraudettes specialize in blaming innocent Actors, rather than the Meddlers responsible, for the existence and the activities of L'il Grabbers.

The Bad GOGsters:

Meddlers. Meddlers run the Public "Works". They enslave Actors to pay for Public Schools, Public Parks, Public Transportation, Public Housing, Public Television, and State Universities. Everything that can be useless to, hurtful of, hard for, and Meddling with the rights of Other Workers is under the control of Meddlers. Meddlers are most dangerous when they try to "help".

Media Frauds & Fraudettes. Because they can Lie *and* Whine with great skill, Media Frauds & Fraudettes receive obscenely high pay for reading simple scripts that tell the other Actors what Public Officials want them to think. They call this process "News".

Media F&Fs" motto: *"We lie. We say we don't."*

Underminers. They weaken Workers better than mere Distractors. For instance: When a Distractor makes a Cowboy Movie, the Good Guy wins and goes off with the girl. The Audience is amused.

When an Underminer makes a Cowboy Movie, the bad guys win and go off with each other while the girl becomes a prostitute to "find herself". Then, some in the Audience have a harder time telling right from wrong.

If it is true or good, innocent or decent, Underminers will tear it down. Underminers, like Distractors, are usually paid indirectly.

Examples: Underminers are given preferential rights to buy newspapers so that they don't fall into the hands of people who would rather tell the truth than lie for Public Officials.

Public Officials reward Film Makers for Undermining values by keeping IRS agents away from their books. They let obedient film makers fraudulently inflate their expenses, which reduces their taxes, but they make them lie to do it. That means they can be arrested any time they try to stop viciously Undermining.

The Most Important Underminers. TV and Radio Stations are rewarded directly. If they Undermine obediently, they are given monopolies of broadcast frequencies. Increasingly, Public Officials take over Direct Control of broadcast media to totally control the information Underlings receive. In a magnificently brazen lie, they call such private monopolies "Public" Television and "Public" Radio.

Another bold lie: Media F&F's, who owe their grossly high-paid jobs to Broadcast Monopolies that keep others from being able to speak freely, will chatter endlessly about "the importance of free speech".

Understanding the depth and frequencies of such lies shows how far above Conventional Reality an Actor has risen.

MAGOG: Maniacally Acquisitive Guilds of Government. Deranged Public Officials cause war or revolution. Then, MAGOG tries to take whole sections of The Stage. MAGOG appears after the GOG has destroyed the individual freedom of their own Workers.

Ruling every Guild of Government
are the two kinds of

PUBLIC OFFICIALS
"Underlings Must Serve to Live "

Visibles. Visibles are either Elected or Appointed. Most Visibles use their status to ac-

cumulate Props and/or Power for themselves
and/or their supporters. A very few think that
is wrong, and are "Principled".

Nearly all Public Officials say that they are
"Principled". But, many of them lie. Public
Officials who lie do so far more skillfully than
mere Media Frauds & Fraudettes.

Public Officials enjoy: 1. Giving Opinions
that Actors have to hear. 2. Making Actors do
things. 3. Being Important.

Public Officials have one power that few other Actors have: they can legally order Underlings to remove other Actors from the stage.

Invisibles. They aren't often seen and heard by other Actors on The Stage. They stay out of the limelight.

They have the ability, working separately or together, to put Visible Public Officials in Office. They can also remove them from Office.

The Best Public Officials. A very few are interested in providing maximum freedom (lower taxes) for the other Actors. They are much beloved by freedom-loving Actors.

Public Officials like Moses, George Washington, and Robert E. Lee assume an important place in the minds of all Actors who love to be free from tyrannical GOGs.

All the Props have been described and you have met the Players.

Each of us can play the Starring Role in our run of *"My Life"* better if we know what the On-stage Directors, the Public Officials, think of us.

The writer of this Playbill has been given improperly obtained portions of the *Public Officials' Handbook,* some of which follow.

Public Officials tend to see reality backwards. The *Public Officials' Handbook* will help you understand that they see you backwards, too.

II

Random Excerpts from Chapter One of The

Public Officials' Handbook

Education. Underlings must be convinced that the Universe is Big, Old, and Complicated. Tell them to study and protect The Whole Universe. *Do not giggle when you say this!*

Each new Discovery about the Stage can, and should, be made into a reason for fear and alarm. Fear and alarm justifies raising taxes to provide jobs for more Underlings.

Schoolies. Force your young Underlings to attend "Schools" under our control. Pretend they will learn things that will help them. Do this because:

1. Underlings must be kept off the street and out of trouble. As Workers' Troupes become more efficient, fewer workers are needed. Make such excess Underlings into Schoolies, from Nursery Schoolies to Grad Schoolies.

2. Underlings must be kept too ignorant to be a threat. Force Schoolies to waste years and years

learning less and less. Those made totally ignorant can be "certified", and allowed to teach.

3. To make Underlings accept the idea that they should do what we say, have certified Meddlers teach them that they are "evolved" from animals. There are many other ways to destroy their dignity, but this is the best beginning.

4. Jobs can be invented so surplus, over-educated Underlings can be given "jobs" to "work" in "schools".

Don't let Educators Educate: We don't want Schoolies who can think and read. We need Underlings we can tax, cheat, and direct.

Underminers must be able to manipulate millions of teachers into pretending that using utterly failed systems actually helps Schoolies learn.

Teach Schoolies to trust your Media.

Always Control Access to Media. We need mechanisms to spew endless lies on command. We need to keep Actors confused about what is going on around them.

Then, they can't effectively fight to be free. If Underminers can teach them that "freedom" means "being selfish", the fools won't want to be free. Media Frauds & Fraudettes should automatically criticize Workers who are not Making, Moving, or Fixing things that directly benefit Public Officials.

They Must Not Lie Too Well. We want Actors to basically know they are being lied to. That way, we remind them that We are stronger than Truth. Media Frauds & Fraudettes should lie just plausibly enough to convince Actors to support causes we've designed to destroy themselves and their children.

Incidentally, Media Frauds & Fraudettes are more dependable when they're deeply in debt.

How to Promote. Underlings who dumb down, steal from, or harm others must have the best chance of promotion. The worst can rapidly rise right through Meddler, past Media F&Fs, through Distractor, beyond Underminer, and on to Public Official.

Underlings Should Fight Amongst Themselves. Meddlers, Media F&F's, Distractors, and Underminers must be made to fight each other viciously, especially when Workers don't produce enough Props to support their growing numbers.

To keep the swelling ranks of the Guild of Government happy, we Public Officials must continually raise taxes on Workers and Civil servants. Troupes in our own GOG will join together and attack us Public Officials if we don't raise enough taxes.

Taxes must always be raised. If Workers protest, bribe their leaders. If that doesn't work, remove them from The Stage.

First Drain Workers. Then, Squeeze Civil Servants. If you tax Workers ruthlessly enough, they will become so helpless that they'll think you are their friend.

Then, weaken Civil Servants. To help drive them into paralyzing poverty, force your horde of Civil Servants to pay three kinds of tax:

1. Tax the Civil Servants' meager salaries.

2. Force the Civil Servants to give additional contributions of time and money to keep us in office.

3. Force the Civil Servants to pay large amounts of money to Union Bosses. Teach them to call such confiscation "dues".

Chapter Two of The *Public Offi-cials" Handbook* has a brilliant section that shows how to:

1st, establish Dribble-down support systems for Tubers.

2nd, turn Workers into Tubers.

3rd, make Tubers into Criminals and put them in jail. Excerpts:

Keep Those Tubers Tubing. Keep them paralyzed by endless lies. Undermine their ambition with endless Freebies. Ruin their minds with Faddy Fears.

Smarter Tubers can be taught to march in circles around the entrances of Dribble Down Dispensaries and trained to face the signs they carry toward the cameras of Media Frauds & Fraudettes while simultaneously chanting "Give us more! Give us more! It's your duty! Give us more!"

That's the most complicated thing any Public Educator should ever teach any Tuber.

Lifetimes Down the Tubes. When their minds are properly deadened, Tubers will stare endlessly into the Tubes which transmit Distractions and Underminings spread by our barely more intelligent, but much better costumed, Media Frauds & Fraudettes.

Smarter Tubers should be trained to use tiny "clickers". With them, they will be able to go rapidly through the tiny number of Public Official-Controlled frequencies that we authorize their Tubes to access.

Importance of "Clickers". The more channels an Actor can see in a short time, the more intelligent an Actor can be turned into a Tuber.

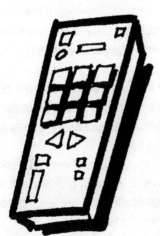

Clickers can intellectually immobilize whole nations. They should be given away if they cannot be sold cheaply enough.

Tuber intelligence may be measured by the speed with which they use their "clickers" to go from one frequency to another. Duller Tubers will watch days and days of one frequency on their huge, multi-colored tube. They will squawk loudly if someone uses a "clicker" between commercials.

Get Good Tubes for Your Tubers. Tubers like tubes that download Lies and Distractions in color. "The color is more lifelike."

Tubes that download in color should be provided free to Tubers. To do this, have Media Frauds and Fraudettes cause "rioting." During "riots", order all Protectors to stay inside until Tuber Teens have transferred all available Tubes from retailers into nearby homes.

Tubers can keep a Tube obtained in this fashion working for four to five years. Since guarantees and warranties do not apply to Tubes obtained by Tubers in the Redistribution Process that ignoramuses call "looting", every few years you will have to arrange another "riot".

When it is necessary for Tube resupply, be out of town. More gullible underlings can easily be made to think that your presence is necessary to keep "riots" from happening.

Don't waste too much money on Tubers. Tubers always vote for us. Media Frauds & Fraudettes should only spend Broadcast time pretending to care about Tubers before elections.

Turning Workers into Tubers. When Workers are free, they're harder to control. So, put them out of work. Destroy free markets, factories, and farms with high taxes and insane environmental regulations.

Remove tariff barriers. Force free people to compete with slave labor available on other sections of The Stage. Reducing disposable income weakens workers. Workers should not have Disposable Income. Remember the First Rule: *"If workers have money, their taxes are too low."*

Invent Pseudo-crimes and Put Tubers in Jail for Committing Them. What better method to build more jails, expand the Criminal Justice system to get more Actors under Direct Control in them, while providing more jobs for Underlings?

The beauty of the Criminal Justice System is that we can tax gullible Workers and Civil servants to pay for the whole thing.

To get more criminals, make more crimes! You can make anything into a crime! Schoolies can be jailed for having snuff! Swatting flies!

If Civil Servants and Workers do things that we say are 'criminal', we *have* to put them in jail. Make sure that they will fear being brutalized by Teen Tubers and real criminals in your jails. Encourage every kind of prison brutality.

Let the Real Criminals *Out* of Jail! Vicious, dangerous criminals don't do you any good while they're wasting away in prison cells. Let them out so they can terrorize citizens.

As you force levels of crime and fear to rise, tell the fools, "We are working hard to reform the Criminal Justice System. We need to raise taxes to hire more policemen and build more jails."

Be Lenient with Rapists, Child Molesters, and Mass Murderers. They're a great asset, but they can be hard to find when you need them, like around budget time. That's when you want them out on the street, doing whatever they do best.

Never staff Parole Boards with victims of crime. Staff them with "Professionals", people who will come up with excuses to return the most dangerous criminals to the streets.

Get the Right Kind of Judges. Be sure to get judges appointed and elected who can say: "We've got to be tough on crime!" at the very same time they can invent hairsplitting excuses to release the very worst kind of criminals from prison.

Use Your Public Schools. They can identify valuable, vicious individuals. Get them addicted to Ritalin and other drugs at taxpayer expense while they're in school.

Be sure that even the youngest children can be confused or stimulated with pornography. Protect child molesters. They help insure future crops of vicious, bitter criminals. Keep enough criminals out there to justify annual three to five percent tax increases. When crime gets out of hand, blame the police.

While doing this, be sure that you live in a safe neighborhood with lots of Protection.

Be Two-Faced. When You Talk to Workers and Civil servants, say: *"We (I) need to raise our (their) taxes to protect (attack) us (them) from (with) vicious criminals!"*

When you tell them this, don't let them see you smirking. When talking to loyal supporters, use the words in parentheses. Sample from *"Speeches you Should Always (never) Make"* from *The Public Officials' Handbook.*

While Creating Enemies at Home, Invent Enemies Far Away. Exaggerate the power and greed of make-believe "enemies" overseas. Workers and Civil servants can be taxed endlessly if threats can be made scary enough.

"War is wonderful. Preparing for War makes us money, Fighting Wars gets rid of troublemakers, excess idealists, and young men who would com-

pete with us if they were alive." Second Rule, The Public Officials' Handbook.

Keep Underlings Under Control. Troupes in the Guild of Government have been known to turn on their Public Officials and destroy them. "It beats working!" Attila told troops of mercenaries when he calculated that Roman Public Officials had spent so much money on games (primitive Tubes) that they couldn't afford decent cavalry.

It's not important that mutinous Huns carried off caravans of inadequately protected Props, raped thousands of inadequately protected women, and sold thousands of inadequately protected children into slavery. The bad thing was that Public Officials were removed from The Stage!

"What anyone does is only bad if it hurts one of us." Third Rule, *The Public Officials' Handbook*

How Public Officials are Ranked. Only a few of the smarter Actors realize that we are ranked by the number of Actors it is possible for us to remove from the stage.

Which Public Officials are Most Powerful? The more power that Public Officials have, the more monochromatic the costumes worn by Actors on the part of The Stage under their control.

Actors are Props. Never fear to remove large numbers of any Actors that can be removed. They are Props. Think of them as Animals. Be sure that you disarm them before you tell them that!

Use Restraint. Don't let Meddlers totally destroy Workers with taxes and regulations. Workers must be allowed just enough freedom and production capacity to produce necessary amounts of Props. *"Never jeopardize Production of Props unless you can completely enslave all Workers."* Fourth Rule, *The Public Officials' Handbook*

Favorite Lies to Tell Frequently. Be fond of saying that "everyone should be free to do what they want." Actors dumb enough to believe us when we say that often die in jail. They deserve to.

On Free Speech. Repeat endlessly: "Everyone has free speech, but no one has the right to yell 'Fire!' in a crowded theater." In actual fact, no Actor should have the right to say anything except that more should be given to Public Officials.

Public Officials need to use Lubricators. *The Public Officials' Handbook* has an entire chapter on dealing with Lubricators. Excerpts from Chapter Three:

Sliding and Gliding. Necessary Regulations should be written by Lubricators. The cleverest Lubricators should work directly for us Public Officials.

Their job is to Slide and Glide increased regulations and taxes by so smoothly that no Actor, Troupe, or Guild will become angry enough to try to physically remove the Public Official(s) responsible from the stage.

Call Important Lubricators "Ambassadors". They work on Deals between Public Officials with the Power to remove lots of actors from The Stage. Call really Big Deals "Treaties" to separate them from agreements that Workers and other Underlings make.

"I've got a good idea! Let's you and him fight." De-Lubricators are skilled at helping us Public Officials cause frictions among factions of Workers and Underlings. Any group can be easily destroyed by being manipulated to go in several directions at once.   Remember the Republican Party?

Holy Oilers Can be Pesky. Holy Oilers tell Actors about the Invisible Producer. They encourage Actors to follow Directions given by the Invisible Producer. They tell Actors what the Invisible Producer wants them to do. Actors tend to trust Holy Oilers. Some Actors freely give Props to Holy Oilers!

Try to corrupt Holy Oilers by getting them to take Props from you.   Have Media Frauds and Fraudettes say that Holy Oilers who won't take Props from Public Officials are "extremists", and "not to be trusted".

If you can, kill or jail them. Develop a cadre of un-Holy Oilers who will tell Workers to do whatever you want.

Holy Oilers: Kill the Best, Buy the Rest. Communists used the Russian Orthodox Church to cement their hold over Workers and Civil servants, but only after they made sure that none of its living Oilers was Holy.

In America, Public Officials have cleverly taken over urban dioceses by funding "Catholic" charities. Sadly, American Public Officials still do not have a means to remove large numbers of truly Holy Oilers from The Stage.

The minute that you get enough power, any Holy Oiler who believes in their ridiculous First Commandment should be exited from The Stage.

What Better Religion Than You? It is a mystery why we, ourselves, are not worshiped. Do not discourage people who suggest this. It is an ideal way to find out which Underlings are the most loyal.

Most Religions have ridiculous superstitions that tell even the most useless Underlings that they have rights. Sometimes, Actors will take these insane notions seriously, and use them as an excuse to keep from doing what you want.

Worse, they might use Religion as an excuse to control or to get rid of *you.* Every force at your command, from Tuber to Underminer, must work to eliminate any serious consideration of religion, aside from the Worship of Us.

Inconvenient Religions. There are still religions that seek to restrict us from removing Actors from The Stage.

Inconvenient Religions can be so extreme that they don't want you to kill unborn Actors, even if you really need to. They just don't understand that if too many of your Actors have children, they may take money you need.

If your Workers get tax deductions for their children, encourage them to get rid of those tax deductions before they're born. That's when they're the easiest to get rid of.

On the richer Sections of The Stage, there are trillions of dollars in additional tax revenue to be gained if you can get those mothers-who-should-be-working to get rid of their unborn children for you.

Any religion that encourages workers to have children is a bad religion.

A Constant Truth: Each Actor knows he or she daily gets closer to the Exit of their Starring Role in "My Life". Don't let them think too much about this, or they may start to think, say, and do things to keep the Invisible Producer from criticizing them harshly at the end of their performance.

One of the most important things your Distractors do is keep Actors from thinking about any notions of Final Judgment. When those stupid Underlings start to think about that, we diminish in importance.

Chapter Four of the *Public Officials' Handbook* Deals With Underling Control. Excerpts:

Work Hard, get Less. An inevitable irony on every section of The Stage: those who work the hardest get paid the least. Don't feel guilty about this. Those who Actually Work aren't supposed to have many Props. The smartest of them realize that having too many Props can cause Swollen Ego, and know how that can bring a Bad Review.

Others pretend to get upset about any appearance of Unequal Prop Distribution. These may form Actors' Unions within the various Guilds. They will hire Lubricators to look out for their interests. Encourage this process.

The heads of such "Unions" will be easily subverted, bribed, and influenced to do your will after their brief "idealistic phase".

Allow Tiny Freedoms. Let Underlings choose their own costumes and hair arrangements. Allow them to paint tiny parts of the Set they live in whatever colors they want. Sometimes, allow them to pick out the food they like to eat. Let them prepare it as they wish. Some may be allowed to own pets.

Workers should not be free to have children. They absorb resources that we need.

Actors like Mobility Props. With them, they can go to places thousands of body lengths away and do things that generate income that we can tax. They'll drive vehicles that we can tax propelled by fuel we can tax. On roads that we can continually repair, close down, and put toll booths on. Progressive Public Officials like to let taxpayers be mobile. Taxpayers call it "commuting". We call it "screw 'em".

You may want Actors to be free to have tax-generating Mobility Props that are wheeled, tracked, or hulled. Winged Mobility Props are available, but should be heavily licensed. We do not want Actors in a position to drop things on us.

Ideally, Actors should be happy with little bicycles, Public Transportation, and walking around. Especially on "Nature Trails" that can, and should, be stolen from private landowners. If you need the billions available from excessive gasoline taxes, then allow them to have automobiles.

Don't Give Them too Much Freedom. Allow Actors to move themselves limited speeds and distances without the permission of Public Officials. They should have to seek, and pay for, your Permission to go from one Section of The Stage to another.

Dole Out Tiny Freedoms. Some Will Think You Are Their Friend. Let Actors seem to be free to select Sounds & Images they want to hear and see.

Only a few are smart enough to realize that the only broadcasts they can easily hear will be on frequencies given to, and controlled by, obedient favorites of us Public Officials.

Younger Actors like to carry Sound Doodads around with them. Let them have and keep sounds on Grooves & Filings. Allow them to view pictures that appear to move. They will take "films" seriously, and talk about them endlessly.

Younger Actors should be encouraged to deaden their minds with discordant, unpleasant, confusing Sounds and Images that Undermine their self-worth. Destroying their self-worth helps keep them from wanting to have, or take care of, their own children.

We Only Want the Right Kind of Parents. Never encourage Workers to have children. It reduces taxes. Tubers, minority groups, and illegal aliens should have lots of children. They are always to be preferred to the children of Working, native-born, majority populations.

Working citizens often think that they have rights. They may raise their children to think so, too. Tubers and Illegal aliens are much more controllable because they know they have no rights. They, and their children, can be made to work for slave wages by clever Public Officials and their friends. Working citizens then have to compete with their low wages. That makes them less likely to have children and forces them to be more dependent on us. Minority votes can be bought more cheaply than the votes of Majority Populations.

Watch out for Uppity Actors. Actors should not be allowed to do Writing or Directing that undermines the Authority of Public Officials.

Distractors focus Actors on useless, preferably self-destructive, activity. If they're too smart to be distracted, use Underminers. Underlings must be kept from shoring up their positions at all costs.

Actors who can't be Distracted or Undermined must be removed from The Stage individually or in large numbers. Sometimes, remove them quietly. Other times, with great fanfare. Sometimes, temporarily. Sometimes, permanently.

Give 'em a Whiff o' the Grape! Outraged Underlings have been known to make substantive changes and get Public Officials who will let them have more Freedom to Direct themselves.

This is very rare. If it looks like it is happening to you, immediately have Protectors attack. Shoot up-

pity Underlings down like dogs, if you have to. That's rarely necessary. Our second-most valuable Prop is the Appearance of Power.

Keep 'em Down & Keep 'em Out. Be ready to act in Brutal Concert to keep mere Actors from having more than paltry piles of Props. Actors must have absolutely no power to do anything prohibited by Public Officials. *"Public Officials are only destroyed by those whom they allowed to rise."* Fifth Rule, *The Public Officials' Handbook.*

Tell 'em They're Free. Then, it's Easier to Tell 'em What to do. Actors who haven't been taught to think for themselves enjoy being directed. Those made unable to use their minds find comfort in having most of their lines written for them.

Self-directing makes such Actors nervous. Even choosing the color and style of their costumes and hair is difficult for many.

All Actors are born with enough intelligence to direct themselves. So, they don't need us unless we can have these skills identified and destroyed by Public Educators. Force young Underlings to spend time with Public Educators.

Meddlers Love to Meddle. Directing the most trivial activities of Workers is unbelievably enjoyable to Meddlers. It's hard to believe that any mind can be small enough to find fulfillment in directing trivialities, but your Meddlers thrive on it.

Meddlers insanely believe that they can improve anything, if only they're given enough Power to Meddle. We don't know what causes them to hate freedom, but they do. They love nothing more than destroying free people and free countries and free markets while telling Workers their favorite false-hood: "We only want to help."

The difference between Meddlers and us? We're smart enough not to believe our own lies.

Chapter Five of The Public Officials' Handbook deals with the importance of Imaginary Problems.

Imaginary Problems: Always a Favorite Tool. Since Imaginary Problems don't exist, no one can ever tell if they've been solved. Never allow an Imaginary Problem to appear to be solved until you have squeezed all possible taxation and regulation out of it.

Any discernable Cause and Effect can be turned into an Imaginary Problem. If you need to blame something, and can't show that it's existence causes something everyone knows is bad, have Distractors invent an imaginary relationship.

Media Frauds & Fraudettes can go on and on about them until you find a new Imaginary Problem for them to babble about. Don't worry about losing credibility. Workers and Civil Servants are raised to have an endless capacity to believe endless lies.

Imaginary Problems and The News. Make sure that every segment of each and every "news" broadcast is a Press Release for some group of donors. *Make very sure that you or your Underlings get paid for publicizing each Press Release.*

It takes prompt action to turn airplane crashes into an excuse to get increased funding for the FAA. Their Union, the airlines, and manufacturers involved will pay you plenty to put the right spin on such "news" stories.

Imaginary Problems Should be Profitable. Asbestos, Alar in apples, carcinogenic cranberries, cyanide in grapes, all these things can be very profitable to someone. People with high-priced replacements for Freon will pay billions to get you to make Freon illegal.

They will even invent more Imaginary Problems, like "ozone holes". That funds academics at State

Universities and other Grant-getters. Everybody profits, except Workers. Don't worry about them. They don't even vote.

The Drug "Problem". Media outlets must convince viewers that Drugs Are a Problem. Magnify these "Problems". Don't remind them that, until our first successful World War, drugs were legal. Today, pain-killers are big money-makers. People mustn't be allowed to kill agonizing pain until we get every cent we can out of them.

Monopolizing pain-killing chemicals in our distribution channels offers more opportunities. Organizations like the FDA are a great help in extorting campaign contributions, and Underlings can be made to think they are "helping".

Keeping drugs illegal makes profit margins so high that even useless illiterates graduated from the Public Schools can be tempted to make big money dealing drugs. When they do, we get to put them in jail. We need crime. We need punishment. We need drugs to be illegal.

Natural Disasters. Encourage construction in areas known to be frequently afflicted with fires, floods, earthquakes, volcanoes, hurricanes, crime, or anything else that repeatedly causes loss of life or property.

Invent bureaucracies to give lots of money to Actors who build in areas where property will be destroyed. Get Underlings used to thinking that people who live in safe areas should be taxed to build houses for those who want to live in unsafe places.

Unnatural Disasters. Not enough Natural Disasters? Burn or blow up buildings to justify more taxing and spending. The FBI and ATF have reliable experts who will know when to burn down or blow up buildings, and will do so without being told or leaving an embarrassing trail. When this happens, go on TV and look "concerned".

If Underlings are killed, look sorrowful, but give commendations and promotions to those involved.

If disasters do not occur naturally, get busy. Why do you think you have Underlings?

Fabricate Bizarre Lunacies. Even Flying Saucers can be used to justify funding for NASA. "A few more research dollars will put widespread fears about cannibalistic space aliens to rest. It's our duty!" Underlings can be told by serious, scowling Media Frauds & Fraudettes on dull days.

The lunacy of NASA reminds people that there is no sane defense against us.

Gun Control. You cannot adequately tax and regulate armed Underlings. They must be disarmed. Media F&Fs should make the private ownership of weapons into a major Imaginary Problem.

To help the fools believe that they should be disarmed, make it look like it's for their own good. Facts should be distorted, twisted, and frequently repeated: "Last year, five hundred children were killed by guns." They should never say "Last week, twenty five thousand unborn children were killed by abortionists."

The Great Unwritten Rule: Each Actor must be given endless opportunities to publicly choose to strain at gnats and swallow camels. Do not remind Actors of that, much less tell them what it means and Who said it.

"THAT'S ALL THERE IS TO LIFE? POWER-CRAZED PUBLIC OFFICIALS ENSLAVING ACTORS ON A STAGE FULL OF PROPS?"

"NO. THERE'S AN INVISI-BLE PRODUCER. HE'S ALSO A CRITIC."

" HOW DO WE KNOW?"

III

Road to Certainty?

A few actors being painfully exited from their Present Performance by Public Officials have said that they have actually seen The Invisible Producer or His assistants during the time of their removal.

Those Actors can be trusted. They are neither influenced by Props nor by Power.

They aren't afraid of Public Officials. They would rather die than lie for or to a Public Official. Their love of truth is why the Public Officials removed them from The Stage.

Is There an Easier Way to Find Out If the Invisible Producer is Real?

The Invisible Producer and His assistants communicate on a frequency that is only audible to receivers that they turn on. If an Actor asks, his or her receiver may be activated. Then, previously inaudible promptings from Offstage may be heard.

All Actors have Receivers that can sense Promptings from the Invisible Producer. Delicate adjustments that allow them to tune in to these wavelengths are made by prayer, fasting, self-sacrifice, and a host of other fine-tunings that any Actor can choose to make.

What Can Actors Know?

The Invisible Producer has set up The Stage so cleverly that it's very hard for the Actors to prove anything about it. Some of them don't even think they can prove they're on it.

There's not an Actor anywhere on The Stage who doesn't have a hard time proving he isn't just a character in another's dream.

He Seems to Have Gone to a Lot of Trouble to Make Uncertainty. Why?

One reason is obvious. If Actors could prove that The Stage was just a stage, they'd be forced to believe in the Invisible Producer. They might be forced to believe the Playbill that The Invisible Producer had written. And, pay more attention to what the Holy Oilers say. If Actors are forced to believe in the Invisible Producer, they won't have Free Will.

Actors who are forced to believe in The Invisible Producer will modify their behavior. Like pickpockets near policemen, many Actors would behave better than they really wanted to.

Uncertainty is what gives Actors Free Will. Greater responsibility goes with greater freedom. Greater responsibility brings greater rewards. And, greater punishments.

Giving Us Free Will was Harder Than Making The Stage

First, He made The Stage and all the Props.

Then, He made the intricacies of the Actors' minds so that they'd have a hard time proving anything, even basic facts, unless He helped or they worked very hard.

At the same time, Actors' minds had to be able to think well enough to get them through big and small decisions every day.

The Actors' minds seem to have been every bit as hard to make as all the rest of Creation.

What An Amazingly Fine Line He Walked to Keep us Free From Certitude

Actors' minds had to be smart enough to Grow, Gather, Mine, Separate, Fix, Make, Move, Lubricate, Pray, Tube, Click, Meddle, Distract, Undermine, and Officiate.

Most importantly, Actors' minds had to be able to know if choices were good or bad.

But, we couldn't be given minds powerful enough to turn thoughts directly into things. The First Actors may have been able to do that to some degree, but all subsequent Actors have had to work to rearrange Props.

?

Is There Utter Certitude Anywhere on The Stage?

Even a confused Actor may say this without fear of contradiction: "I am Utterly Certain that it is possible to believe with Utter Certitude that we may have been made in such a way that we could not have Utter Certitude."

What the heck kind of Utter Certitude is that? It's a first step to the kind of Certitude that lets us know about the Invisible Producer.

So, Just How Well do the Minds of the Actors Work?

The Actors' minds work well enough to consider that Actual Knowledge may be theoretically possible. If they choose, they can move beyond that and learn things about the **I. P.** by studying the set along with past and present performers.

As Actors get close to their Exit, they often choose to think very, very clearly about the **I**nvisible **P**roducer. There is a Persistent Rumor that they are given a Quick Preview of Playback. They see every Direction they failed to follow during their performance. To quote The Rumor: "Their lives flash before their eyes."

Immediately after this Quick Preview, Actors are said to have one final chance to acknowledge the The **I**nvisible **P**roducer and their remorse for not following His directions.

These final thoughts are said to have a quality of clarity that is absolutely frightening.

Another persistent tale: if Actors in the few moments before their Exit sincerely tell the **I. P.** that they are truly sorry for not obeying His Directions, He may reduce the punishment they'd have otherwise received.

Warning: Some Holy Oilers believe that a last minute request for forgiveness following a lifetime of wilful sin may be ignored. Others believe it will be granted. All Holy Oilers agree that minimizing sin reduces risk.

"Payback" follows "Playback"

The Final Criticism may be preceded by "Playback" and followed by "Payback". In "Playback", everything the Actor did is downloaded and played back to the Actor's ether disc (soul) in fast forward.

"Playback" may also show all the consequences of everything the Actor chose to do and not do. After "Playback", there's no doubt as to the validity of The Critic's decision.

"Playback" and "Payback" make the notion of forgiveness especially attractive. Actors know that without forgiveness, no Performance is good enough.

"Did I do that?"

What Role do the Smartest Actors Choose to Play?

The Most Intelligent Actors try to find out things about the Invisible Producer. Oddly, because of the difficult work they did, we know more about the nature of the Invisible Producer, Whom no living Actor has seen, than we know about the nature of the water that we drink every day.

At various times, He gives, and has given, Utter Certitude to various Actors. He would often instruct them to pass His Words along to their neighbors. Most were mocked and ridiculed for sharing the Utter Certitude they had heard. "You are ridiculous!" other Actors would tell them, with Utter Certitude.

It looks like He went to an awful lot of work just to help a very few people. If He didn't think He was getting a good return on investment, He wouldn't have done it.

Tentative Conclusions

1. The **I**nvisible **P**roducer is scouting for Actors with the talent to refrain from doing wrong, if only because they're smart enough to know they may be caught and fittingly punished.

2. The Stage is set so that Actors can freely choose to do wrong any time they want, sometimes hundreds of times a day.

3. The **I. P.** is neither impressed with Media Frauds & Fraudettes nor with people who believe them. He does not like Meddlers. He is not amused by Distractors. He positively hates Underminers.

4. No matter how important, Actors, even Public Officials, are at their Status Levels only because the **I.P.** wants them there.

5. The **I**nvisible **P**roducer has provided many, many things that Actors can freely choose to think are more important than Him.

The Talent for Which He's Always Scouting?

He looks for Actors who Freely Choose to look for Him. He's always hoping that Actors will move toward taking the steps that lead to seeing that The Stage may simply be a stage, and that they are merely Actors.

When an Actor reaches that level of comprehension, the First Step, he or she may climb up to the Second Step: looking for and finding the Invisible Producer.

The Third Step: First learning about, and getting a sense of the overpowering Power that the **I**nvisible **P**roducer has and obeying Him out of quivering, tremulous fear.

The Fourth? Falling completely in love with Him.

At any given time, an Actor is close to the next step on the stairway that will take him closer to the **I**nvisible **P**roducer.

4. FALLING IN LOVE WITH GOD

3. LEARNING TO FEAR GOD

2. I'M GOING TO SEE ABOUT GOD

1. MAYBE THERE'S MORE TO LIFE

stairway for the stars, to the stars

One Approach to the First Step

All Actors pile up Assumptions in their minds. From the top of the pile, they look down to see if their assumptions are neatly and tidily stacked. Then, they consider the view.

Many reach this Conclusion: "If there is a God, He can give more than money. He can give me perfect joy on This Stage, and provide an absolutely wonderful role for me in The Next Performance That Runs Forever."

Actors Direct Themselves

At any moment, each and every Actor on The Stage chooses which way to move. An Actor can only move in two directions: He can move toward the Invisible Producer or away from the Invisible Producer.

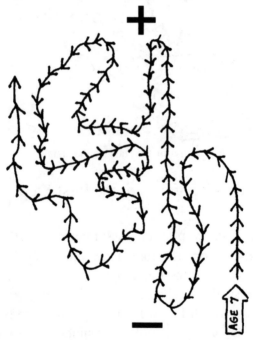

One way to see where you're going is to look down on yourself.

Two Kinds of Prompters

During the starring role of "My Life", each Actor gets conflicting cues and confusing prompts from both onstage and off. Visible and Invisible prompters are always whispering prompts. The cues he chooses to follow will either move the Actor toward or away from the Invisible Producer.

Onstage, Bad Actors guide Players toward sin, slavery and death. Offstage, Demons are doing the same.

Actors Are Given Just Enough Time to Make Their Moves

Each Actor has enough time On Stage to choose and play a Role. Each Role is long enough for The Critic to pass judgment on how well each Actor chose and played the part.

It's easy to understand why Shakespeare called his theatre The Globe.

IV

Audiences

One Audience is On Stage.
Two Audiences are Off.

A ctors are keenly aware of the Actors near them on The Stage. A few are dimly aware that their performance is being observed, criticized, applauded, and influenced by Audiences Offstage.

Actors are Observed by
Three Audiences

Each Star of "My Life" is aware of nearby Actors on The Stage. At each moment, that Audience is composed of all the Actors who are considering the Star's performance.

There are two Invisible Audiences that must be considered if an Actor is to minimize the risk of A Bad Review after his or her starring role in "My Life" is over. A way to picture the Offstage Audiences:

The Invisible Audience Above The Stage:

1. The Invisible Producer
2. His Immediate Family
3. Stagehands and Crew
4. The Souls of Actors who chose to Act as if they wanted to be Saved

The Invisible Audience Below The Stage:

1. The Devil
2. His Demons
3. Souls of Actors who chose not to Act as if they wanted to be saved

The Offstage Audiences

Members of the Offstage Audience are allowed to prompt Onstage Actors, but they are not allowed to be obvious. They may be allowed to rearrange thought particles to get an Actor Onstage to reach different conclusions and change Directions. When that happens, an Actor may say: "I've got an idea!"

Some Prompters may take things from printed pages, guide them into Actors minds, and influence them. Others ride wavelengths like cowboys astride flying snakes. They can ride an idea right through the cornea, zip through the clear, vitreous humor inside the eyeball, and zoom right through the retina, down the fiber-optic nerve, right into a receptive area of the Central Processing Unit.

Actors Can Communicate Offstage

Each Actor on The Stage owes a debt of gratitude to individual Troupers who preceded them in previous performances. Many earlier Actors purposely left Props to Actors they knew would need them. Some Actors have enough sense to be grateful.

Actors' Most Important Legacies

All Actors leave arrangements of teeny-weeny particles (thoughts) in the minds of other Actors. These can be the most valuable Doodads of all. The best Actors spend the most time trying to Direct their own thoughts, words, and deeds in accordance with the desires of the Invisible Producer.

Before going Offstage, if an Actor shares his knowledge with a younger Actor, the younger Actor has a very real advantage. He can reach better Conclusions more quickly.

A seasoned Actor who tells a younger: *"Try to never give the Invisible Producer a reason to be angry with you."* will help lead that Actor to a lifetime of healthier conclusions than someone who passes on something silly like *"The person who dies with the most Props wins."*

Words and Examples that help a person get to the Right Conclusion are the finest legacies that can be left.

An Actor who inherits such a valuable legacy can get a better seat for the exited Actor who gave the good advice while Onstage. More amazing, the Actor Offstage is then in a better position to help Favored Performers in the Current Performance.

Favors passed between the Actors on Stage in the Present Performance and Actors who've moved Offstage is like people tossing a diamond right through a wall and then getting it thrown back.

The diamond gets bigger and brighter each time it passes from Offstage to On. And, no one can even prove it's there.

A Way to Visualize Stage & Audience:

Imagine that The Globe is empty. Then, picture an Actor wandering around on it. Then, two. Then, a family, followed by a clan. Then, a tribe. After that, a nation. Then, an Empire. Then, the Empire breaks apart into nations, peoples, tribes, clans, and families.

The Invisible Producer left a description of the Whole Production. The first part, "Genesis", describes what actually happened in Iron

Age words that we may want to translate into
Information Age equivalents to understand
them better. (See "My Yoke is Easy and My Burden is
Light." Old Drum Publ., 1997)

Actors can see The Stage, its history, and
their own performance better when they visualize from an Offstage position. Every Actor
is made with the ability to be able to do this.

Picture a Very Vacant Lot

The Playbill says that at first, the Construction Site "was without form and void".

Then, the Producer built The Stage. You
know what The Playbill says. The Stage was
built in seven steps. It had all the Basic Props,
but only one Actor.

After nearly a week of Set Preparation, an
actor was put on the Stage. He was Bo-o-o-o-
ring. The Offstage Audiences began to lose
interest. "A week's work, just to watch a
naked Actor give names to Props?", muttered
a few of the Stagehands who'd joined the first
Union run by goons.

It Didn't Stay Empty for Long

To liven things up for the Offstage Audience, The Invisible Producer put a new type of Actor on The Stage. An Actress! Catcalls from the stagehands. The Offstage Audiences perked up. The two Actors were able to bio-fractalize Other Actors. They did. One Actor killed another! The Offstage Audiences were riveted to the performance. Things were getting interesting! Free Will was something to see!

Boiling Down the
Histories of The Stage

More and more sets were added to more and more sections of The Stage. More Actors produced more Actors. Some Grew and Gathered. Others Mined and Sorted. Some Moved and Stored. Some specialized in fires. Some, Doodads. Actors and Actresses bustled around the Stage and filled it with activity. They burrowed and dug, planted and reaped, forged and sharpened, spun and wove. Busy, busy, busy.

Some built dams. A few miles away, others drained lakes. Some planted trees. Others cut down forests. Always doing something.

Good Actors Want to do Good

Some of the Actors remembered, some figured out, and some were told directly that there was an offstage Production Unit that included an Invisible Producer, an Unseen Director, and the greatest Writer of Creation.

Good Actors tried to figure out what the Three Invisible wanted them to do. Later on, a few of the Actors were so very, very good that they were allowed to discover that the Three Invisibles were actually different aspects of One Being.

The Good Actors tried to do what they believed the One Invisibles wanted them to do. They tried to find out the best Characteristics that they could choose. The Bad Actors hated any directions but their own. Still do.

Good Actors

Having lots of Props can be fun for a little while. Good Actors want everyone who wants a lot of Props to have lots of Props. Owning too many Props is silly, and a waste of time. Actors have an easier time figuring that out if they actually go through that stage. People who have lots of Props have a great opportunity to move to a higher level of The Stage. If they want, they may have the fun of finding out that Props are just Props. Then, they get to move into the far deeper joys that can be reached after realizing that having lots of Props is childish.

The smartest Rich People give their Props away to people who don't have a lot of Props and who think that the very best way to use Props is to bring new Actors (children) onto The Stage.

After they've given away the Props that used to own them, they go off, somewhere and climb. Not mountains. Ideas. They think and talk and read about the Invisible Producer. They talk to Him. And, listen to Him.

Bad Actors just hate it when someone they know does that.

V

Universal Constants

No matter what Guild or Troupe they are in, all Actors spend their time on Stage doing just four things. Each Actor at each moment is either:

1. Giving or Getting Opinions

2. Manipulating Props

3. Rehearsing

4. Sleeping. While Sleeping, an Actor often dreams about the other three Roles.

Giving Opinions

As soon as someone takes on the role of Opinion Giver, they announce it to the other Actors by prefacing their very next comment with words like "I think. . .", or "It seems to me . . ." At that point, well-mannered Actors within earshot magically are transformed into Opinion Getter, or Audience.

While in the Audience Role, the Actors judge the performance of the Opinion Giver. They judge on What, Why, and How the Opinion was given.

Many in the Audience give opinions to the Opinion Giver while he is giving his to them. They do this by scowling, frowning, nodding, smiling, etc. Sometimes, Opinion Givers notice this, and modify their opinions while they are giving them.

As soon as the Opinion Giver is finished, and he is sometimes hurried along by murmurs from the Chorus part of the Audience, "Get on with it, get on with it, get on with it!", many of the other Actors stop being Audience and try to become Opinion Givers.

Usually, an Actor emerges from the Audience to take on the starring role of Opinion Giver by saying, often politely: "You are obviously smart/dumb, right/wrong, knowledge-able/ignorant. I think The Correct Opinion on this subject is the same as yours, different from yours, partly similar, etc., ..."

Why do Actors Enjoy
Giving Opinions?

Actors who Give Opinions effectively can influence other Actors, and make their opinions widely shared.

Actors love to Give Opinions. Some actors have to practically be restrained from excessive Opinion Giving.

Some Actors give opinions by asking questions. "Why does So-and-So spend so much time giving opinions?" is a clever way that Actors have of giving the opinion that So-and-So spends too much time giving opinions while pretending to be humbly asking another to Give an Opinion.

Children go to School to Learn to
Give and Evaluate Opinions

Since even before St. Augustine was teaching Public Speaking, Actors have had schools where students could learn to give, get, and analyze Opinions.

Actors who learn the most accurate Opinions about Props and Actors will be better able to figure out new ways to deal effectively with them.

Players learn to channel electric charges more productively between memory particles (thinking) by having erroneous connections corrected by more intelligent, and/or by more knowledgeable Opinion Givers.

Schools that help students provide important benefits. Schools run for the benefit of Public Officials are worse than useless.

Debates, Discussions, Arguments

Sometimes, Actors enjoy Impromptu Performances where two or more of them quickly write and exchange Opinions. An Actor begins such a Playlet by Giving an Opinion. Then, he plays the role of Pretending to Be Willing to Accept Correction while he listens to an Opinion about his Opinion.

After that, he Rewords the Original Opinion to try to get it to Stand. Then, he may listen to re-worked Opinions as to why he is still not completely right. After getting Still Other Opinions about his Opinions, he refines and re-refines those Opinions until no one can maintain interest or improve his Stand.

The goal of all this? Each Star in "My Life" wants To have his Opinion Stand.

How do Actors Know Whose Opinion was Correct?

Usually, the last Actor making points is judged to be correct. He may announce that as a fact. "Well, I guess I was right, wasn't I?" Usually, other Actors won't admit that, and often respond: "You aren't right, you're just rude and overbearing."

What they have done is to change the Critical Standards from logic and facts to mushy feelings.

No Opinion Giver worth his salt will refrain from mentioning that.

What's the Point of Giving Opinions?

Actors Give Opinions to get other Actors to do or think things that will benefit either the Opinion Giver or some cause or group favored by the Opinion Giver.

Lurking behind every "I think" is a "you should".

Opinions may be Based on Logic.

Opinion Givers who think that reason and logic are the most important guides to opinions tend to say things like "Let's be reasonable".

They may believe reason is the way to settle differences. They also know that if they are able to say something that seems Really True, no other actor will be able to argue logically with them. Then, they can maintain the coveted role of Opinion Giver for a longer time.

Let's be reasonable

Opinions may be Based on History

Lubricators, particularly lawyers, refer to documents that recorded Widely Accepted Opinions of Past Performances. Magna Carta is an example. So is The Constitution.

"My opinion is the same as that held by the Framers of the Constitution. That was right then and it is just as right today." a legislator or jurist may say.

Jewish Opinion Givers frequently refer to the opinions of Moses. Most of the time, faithful Jewish people will defer to his opinions if they differ from their own.

Sincere Christians try to base their thoughts, words, and deeds on what they think Jesus would have wanted them to do. Other Actors sneeringly call these people "The Christian Right", in an attempt to invalidate the credibility of any Opinions that negatively affect their own Opinions or desires.

Opinions may be Based on Feelings

Some Actors are so obsessed with getting more Props and Power that they ignore reason, logic, or Historical Precedents. When they Give Opinions, they begin "We (I) want . . . .", or "We (I) feel. . ."

Politically, these Actors tend to think of themselves as "Progressive", a polite term meaning "thief ". When "Progressives" become Public Officials, they tend to turn Protectors into heavily armed Grabbers and sic 'em on the helpless Workers they're always trying to "help" by having them disarmed and enslaved.

They lie and steal and kill because that's how they want to get what they want. They only want. They are obvious because they always lie.

Actors Base Their Opinions on What They Think is Best

Some Actors want Props more than anything. Others think the best thing is to get a Better Role in the Future Performance. An Actor who takes a Vow of Poverty will Give Opinions that differ widely from those obsessed by Prop Accumulation.

Few, if any, Holy Oilers have thought the Invisible Producer was impressed by Accumulating Props.

Assumptions lead to Conclusions; Strong Opinions About Conclusions May Lead to Action.

After an Actor puts two or more Conclusions together, he may Act.

An everyday example of an Actor putting Conclusions together: "I want more Props. I need them. To get them, I have to trade my time and talent for more than I earn now. I must act. If I don't get a raise, I will try out for another role (job) that will give me more Symbols for Props (take-home pay)."

Why do Some Actors Want More Props and/or Power?

Actors can be happy with what they are, have, and can be. The Playbill provided by the **I**nvisible **P**roducer suggests that each Actor has all that is necessary. It further states that each Actor should thank the **I**nvisible **P**roducer for his role and for all that has been given to play it.

Most Actors do not automatically want much more than they have. Most Actors do automatically want as much, or slightly more than, the Actors around them.

Having Props a/o Power may have an effect on the Standing of Opinions. "Well, he is rich and powerful, so maybe he's right." is a common response when Actors hear a rich, powerful person Giving an Opinion.

"He is rich and powerful, so maybe he's out of touch with what's going on." is also a common response when Actors hear a rich, powerful person Giving an Opinion.

Enough Props and/or Power will shut down objections to almost anything. When a Public Official with enough Power to remove whole continents full of Actors from the Stage shows other Public Officials pictures of his newest Death Doodads, they may be less likely to argue with his Opinions and Conclusions, much less stand in the way of his Desires.

VI

WAYS TO RUN THE STAGE

Monarchy: An Efficient
Way to Run the Stage

Before Democracy, there were Rulers and Monarchs. Such Public Officials hold Office by inheritance or appointment.

Monarchs are what Rulers call themselves after the First Generation. In earlier performances, such Public Officials were presumed to own entire sections of the Stage, including Actors. When they acted as if they owned the Actors, they tended to take good care of them.

Most Actors didn't mind pretending that the King owned them. They were taxed at low rates, and the King used most of the tax money to benefit them. He fixed up castles for defense, trained Protectors, and piled up valuable Props. Good Kings wouldn't let many Meddlers go around Undermining and taxing the other Actors. Most Kings took better care of Workers than Public Officials today. Workers only look fondly on Public Officials because their ancestors enjoyed life so much during the periods when Public Officials actually had policies designed to help Workers.

Most Monarchs didn't take themselves as seriously as they pretended to. Today, most Monarchs just have a few tarnished Props.

Monarchy: Too Efficient

Peaceful kingdoms were filled with law-abiding people. A good monarchy didn't provide Actors with opportunities to exercise their Free Will by choosing to lie or steal.

So, the Invisible Producer, working behind the scenes, let some of the Smarter Actors try out various kinds of Tyrannies and Committees as they developed the concepts of Democracy.

Democracies provide more opportunities for everyone to lie and steal. At the Final Criticism, many Actors wish they hadn't lived in a Democracy. "In a Monarchy, I wouldn't have had so many chances to lie and steal." they cry, just before the trapdoor opens.

Democracy: Gloriously Inefficient

On a Democratic section of The Stage with 10,000 Actors, a Public Official does *not* need 5001 votes to be able to take Props away from Actors (raise taxes).

Out of 10,000 people, only 6,000 are eligible to vote. Only 33% of those eligible, 2,000, actually bother to vote in many Primaries. About half are in each party. Special Interests can *often* get what they want if half of *one* party's primary voters (501) vote their way. They can *always* get what they want if half of *both* party's primary voters vote their way.

A Public Official only has to take a little money from 9,459 people and give it to 500 party hacks to get their votes. Their vote, plus his, may let him Rule the Democracy.

Democracies work this way when they are taken over by people who want to profit from public service more than to be right.

Democracy's weakness? You *can* fool (or buy) some of the people *all* of the time. Only a tenth of the residents need to be fooled (or bought) in Primaries to give Socialists *all* the candidates in both parties in General elections.

Broken Democracies can be Fixed

Even though Public Officials have power, they have to at least pretend to respect the other Actors. Those who don't respect the Actors can sometimes be kept from Directing. If Public Officials abuse Actors with excessive taxes and regulations, they may be driven from office.

Many Public Officials in a Democracy truly respect the other Actors. Many do not. Many work hard to disarm Actors and turn them into slaves.

Why Actors Like Democracy

Every Actor believes that His Opinions have value. People who give lots of Opinions consider themselves to be quite generous. Such Actors hated Monarchy because it automatically reduced their own self-importance. Bejewelled and enthroned, Monarchs tended to monopolize Opinion Giving. Democracy lets more Actors Give Opinions.

Many Actors enjoy being asked their opinions by the many pollsters used by Public Officials in Democracies. "I told him. . ." Actors enjoy saying.

Most Actors think that they should be asked questions by pollsters more often. Bad feelings about Democracies would disappear if Actors living in them were polled more often and if those opinions were not ignored by self-serving Bureaucrats.

Hmmmmm

Actors believe that they are given the right to vote because their opinions actually are valuable.

If Actors' opinions are truly valued and respected by Public Officials, why do Public Officials ignore them and try to change them?

Jury Duty—Bulwark of Democracy

Jury Duty makes Giving Opinions very important to all the Actors involved. Every single Actor who has been on Jury Duty says "And, then I said . . . ." about any number of

Opinions they Gave. They may complain about being on Jury Duty, but they love it.

Juries give jurors the opportunity to say for decades, "And, then I said . . ."

The Jury, a Brilliant Invention

The opportunity to talk for years about "Important Opinions That I Have Given" makes all Actors who are given jury roles grateful to the Public Officials who went to all the trouble of giving them a chance to Give Big Opinions.

Public Officials make frequent references to "the vital importance of serving on a Jury." All who have served on a Jury enjoy being recognized, even indirectly, by Public Officials. Barely one Actor in ten thousand will notice how easily they are manipulated by such flattery.

Actors who've Given Opinions on Juries have the self-confidence to Give Opinions more freely than those who have not participated in Important Opinion Giving ceremonies.

Former Public Officials are almost as impossible to keep quiet as former Jurors when Actors trade stories of "Important Opinions I have Given".

Even Shy Actors Like Juries

Some Actors will not register to vote because they do not want to have to serve on a Jury. "I don't register to vote because I don't want to serve on a jury. I just wouldn't feel right about giving My Opinion in a trial." is their opinion. And, they enjoy giving it.

Actors Not on Juries Like Juries, Too

Actors enjoy analyzing the verdicts that other Actors give. "Boy, they sure were stupid!" is an Opinion Given by many who then say: "I wouldn't have voted that way. I would have . . ." So, the Jury System gives them, too, an opportunity to participate by Giving Opinions on Opinions.

VII

Notes on Tubers from the Public Officials' Handbook.

Public Officials are fascinated by Tubers. Tuber research is endless and ongoing.

The *Public Officials' Handbook (Democracy Edition)* has an appendix that records miscellaneous Tuber Observations. Excerpts follow:

Tuber Pretensions. Few Tubers will admit that they don't have the slightest idea about what's going on around them. It is un-heard of to hear a Tuber say: "I have been a simple Tuber for forty years. I do not have an intelligent notion of anything more complicated than what I like to eat, tube, and wear."

Who's a Tuber? You're a Tuber. I'm a Tuber, too. Regardless of their Role, when Actors Tube, they drop to the level of the Tubers. Even Public Officials have been turned into Tubers by too much Tubing.

Tubers can't tell Democracy from Television. Television, combined with Democracy, gives even the most immobile Tubers the notion that their vaguest whimsy is so valuable that it should be shared with large numbers of people for long periods of time. The louder and longer that the fatuous Big-Mouth Tubers get to share their opinions, the happier they are.

The Joy of Giving Opinions is quantified by the Happy Blabber Number. It is calculated with the simple Happy Blabber formula: Multiply the number of seconds that a Big-Mouth Tuber spends Giving Opinions by the number of Actors who will hear the Opinion times the decibel rating of the Big-Mouth Tuber.

No Happy Blabber Number is high enough to either satisfy or embarrass a truly Big-Mouth Tuber.

Democracy + Television = Tubervision, Form Without Substance. Tubers love the opportunities to Give Opinions that Televised Democracy provides. They believe that Being Heard is the same as Having Power.

Few Actors are as happy as a Tuber on a Talk Show. They are practically transfixed with joy when their feelings are broadcast. Talk Shows should follow this proven formula:

Wide-eyed, Breathless Host: "And how did you *feel* about that?"

Tuber, grim-faced, nodding: "I felt that it was *really bad!*"

Breathless Host, moving close to Tuber: "And, what did you *think* about *your feeling*?"

Tuber, staring right into Camera a few inches away, "I *think* that my *feeling* was *right!*"

Tubers in the studio audience, each of whom would be equally able to make the same noises on cue, will explode into applause. They are *not* applauding the Tuber on the Tube. They are applauding themselves because those are the very same lines that they have learned and would have given with just as much *feeling.*

They, too, could have made, and talked about, the same rapid journey from Feeling to Thinking About Their Feeling to Moral Superiority.

Television Makes Many Minds One. Yours. Millions at Remote Tubing Sites (dwellings) do not participate in such shows by clapping their hands, but they do nod vigorously. "That's right!", they say to themselves at appropriate times, sinking into the mindlessness. Each time they silently say "I would have said the Very Same Thing.", a tiny piece of their individuality disappears. Reduce their ability to think independently in tiny ways every day.

Tubers derive deep satisfaction from saying out loud to themselves, and to nearby Actors, "That's just what I would have said!" Most rehearse exactly how they will talk about their feelings if they ever get a turn to Give Opinions on the Tube.

Even more amazing: Many Tubers will listen carefully to Media F&Fs, memorize what they say, and repeat the cunning lies as if they were giving their own opinion. Public Schools must teach that mindlessly repeating grotesque Tube lies is sure proof of High Intelligence.

The total lunacy is so great that the Media Frauds & Fraudettes to whom millions of Tubers spend hours intently rehearsing the regurgitation of their own lies neither know nor care that the particular Tuber is even on Stage.

Use "News" to divide The Stage into Common Enemies and "Our Friends". Tubers are fond of memorizing, rehearsing, and repeating Opinions about Props, Actors, Troupes, and Guilds whom

Media Frauds & Fraudettes have convinced them to consider either "enemies" or "friends".

The manufacturing of "common enemies" and "our friends" is the most important job done by Media Frauds & Fraudettes. Within days, highly skilled Media F&Fs can convince Stages full of Tubers that entire groups of Props or Actors are "enemies". What the Media Frauds & Fraudettes call "News" is, of course, just an update of how we Public Officials are making, sustaining, and changing our list of "enemies".

Remember, the "News" must never be more than an up-to-the-minute roll call of who or what has to be watched, controlled, stopped, rewarded, or squeezed.

Be sure that "Enemies" have the opportunity to become "Friends" by giving money to those who control Broadcasts of ink or electrons. "Friends" must become "Enemies" if they don't give enough money to the very same people.

Meddlers should be trained to call this process "Public Relations". In earlier performances, it was called "bribery", or "payola".

Tubers should be kept from thinking about this process. Train Young Schoolies to wave, yell, and act like fools when an opportunity to be Tubed presents itself to them. Force TV weathermen to visit elementary schoolchildren. That's a good way to train young Actors to make fools of themselves near a Media "personality".

Keep 'em Close. On each News Show, huge, magnified faces of grossly overpaid Media Frauds & Fraudettes should be stuck right up close to the Tubers. They must look very solemn when they announce who the latest "common enemy" is. Typically:

"Uncontrolled Private Sector interests are widely believed to be responsible for newly discovered cancer-causing chemicals endangering us and our environment.

"Concerned Public Officials believe that the problems can only be solved by more stringent regulations and increased site inspection.

"It is our duty to protect the environment from those who would destroy it for personal gain."

Tubers will frown and remember that the Private Sector is bad, greedy, and must be undermined.

Closeness = Trustworthiness. If the faces of the Media Frauds & Fraudettes *look* close, Tubers' instincts force them to think that Media Frauds & Fraudettes actually *are* close. Many Actors talk to, and Give Opinions to, Media Frauds & Fraudettes who star on such shows, and to whom they feel, well, very close.

Be sure Broadcasters show the same Media Frauds & Fraudettes up close, over and over again. Tubers become familiar with, and more trusting of, the close-up faces of the Media Frauds & Fraudettes than members of their own families. Tubers will believe a Media Fraud who's done nothing but lie to them for half their life rather than believe what their own father says is right.

Media Frauds & Fraudettes must be continually trained to look honest by people who spend their careers studying ways to look honest. Keep it so that Tubers can't help but trust Media Frauds & Fraudettes.

Tubers are so Trusting That We Must Work to Despise Them. Tubers persist in believing that our Media F&Fs care deeply about them and about important issues. Tubers will write letters to them. Those letters should be answered. Mechanized word processors keep such expenses down.

Even more bizarre: When a Media Fraud or Fraudette becomes incompetent, can no longer lie effectively, and has to be de-tubed, enraged Tubers will call The Broadcaster and complain. "So and So shouldn't lose that job! I could always *trust* So and So!"

When Tubing, the Brain-rank of any Actor drops precipitously. Unless he is watching "Jeopardy".

Media Frauds and Fraudettes are Brighter Than Tubers. Barely. Media Frauds and Fraudettes need one or two well-developed skills. First of all, each one must be able to arrange its hair in a distinctive, easily recognizable fashion.

You must also require each Media Fraud and Fraudette to develop characteristic and obvious signals to convey concern to watching Tubers.

Jerks and Twitches. Local Media F&F's often have to tell the truth because local events may be verifiable. They should be adept at a minimum of two of the following: Lip-pursing, rapid head movements, sphinx-like gerbil stares, subtle, near-spastic jerks and twitches, brow-furrowing, forehead wrinkling, scowling, eye-narrowing, shrugging, supercilious smirking, eyebrow-raising, hearty chuckles, and wide-eyed astonishment.

Network Frauds and Fraudettes must be adept at *all* of the above. They must also be able to do appropriate voice intonations while they chuckle, twitch, frown, smile knowingly, etc.

Hand gestures are newly popular. Polled Tubers think that Media Frauds and Fraudettes who make hand gestures are sincere and concerned.

All of your Media F&Fs must be able to remember that anyone who does anything that could deprive Public Officials of more power must be made to look like liars, losers, or 'extremists'.

Warnings: MF&Fs must be discharged when they can no longer keep their eyes wide open while they are lying. Be sure your Media Frauds and Fraudettes don't use transparent eyelid props. Have dressing rooms inspected for eyelid props and fire whomever uses them. Lies are far more believable when they come from people who look fresh and spontaneous.

Broadcast executives, directors, and writers all sneer at the intellectual pretensions of the Media Frauds & Fraudettes. Do not let them do so publicly. The only thing that lets Media Frauds & Fraudettes lie effectively is credibility.

Do not cut corners by trying to have only one or two fast-moving Media Frauds or Fraudettes on the "News" who try to appear to be more than one or two people by appearing in different hairpieces and costumes as he or she moves from "News"   to weather to sports. Even the dumbest Tubers will sense that something is wrong. Check frequently to see that it isn't going on.

You should have at least two Frauds and two Fraudettes on each "news" show, preferably of different races.

Each should be skilled in different types of jerks and twitches. They should be able to chuckle knowingly with absolutely no discernible provocation.

Encourage them to carry on inane conversations with each other. Train them to laugh believably at even the stupidest jokes. This makes Tubers think that being stupid is genuinely funny, and that's how they will try to be funny.

VIII

The Lottery Mentality

What The Public Officials' Handbook Won't Tell:

The Real Reason Media Frauds & Fraudettes get Paid *so Much Money*?

The Great Mystery: Why are Media Frauds & Fraudettes paid so vastly much more than their fourth grade reading skills are worth?

Certainly, they lie beautifully and endlessly. But, Liars, even skilled ones, may be hired inexpensively. Media Frauds and Fraudettes do something else that's vital to the survival of Public Officials. They reinforce the Importance of Each Tuber.

Each and every Tuber can look at a well-trained Media Fraud or Fraudette making five million dollars a year reading lies in fourth-grade words, and say in perfect, undeniable truth: "I could do that." The logical implication: Each home Tuber can, and does, actually feel as if he or she is worth five million dollars a year.

Broadcasting proof that every single Tuber is important helps keep them quiet. By watching mindless, overdressed, lying nincompoops who can barely read one syllable words out loud, every single Tuber can honestly believe that he or she has the ability, if not the right, to make five million dollars a year.

Most Actors on The Stage don't make nearly that much money. They conclude that's because no one has "discovered" them. In their heart, every single Actor truly believes that he or she is just an "accidental discovery" away from being a highly paid "personality" on The "News".

This is why Media Frauds & Fraudettes are of every sex and nationality. Tubers who vote or might vote must be provided with a way to identify with even the remotest chance of getting a five million dollar salary, even if they had suffered gross brain damage.

If Cocker Spaniels could vote, there would be Cocker Spaniels coifed and costumed so well as to be barely distinguishable from the other Media Frauds & Fraudettes, yapping about what the "News" is.

The Lottery Mentality/Hope Springs Eternal Disorder

The Lottery Mentality, "Make 'em Think They Have A Good Chance of Winning, Even If The Odds Are Ten Billion to One Against 'em", is the result of perverting something that The Invisible Producer programmed within all Actors' minds. To the ancient Greeks, Hope was what was left in Pandora's Box.

Originally, Hope was a healthy thing that gave each Actor the notion that the next audition could be successful. Public Officials pervert Hope into a Logic-destroying Disorder that keeps Actors from thinking clearly.

For example, Actors are subtly taught to think that if there is one survivor on an airplane wreck that "It wasn't so bad, after all." because each Actor has been taught math and logic so inadequately he thinks there would have been "a good chance" that he would have been the person who survived if he'd been on the plane.

Public Officials foist the Lottery Mentality off on Actors so well that even Math teachers in their Public Schools will buy Lottery Tickets!

Public Education has raised younger Actors' self-esteem to such lunatic levels that they "feel good about themselves" even as they stand in long lottery lines to willingly participate in their own fiscal destruction.

After Lottery-Think, Enslavement

Public Officials have manipulated the Hope Instinct so that the hundred million to one odds of winning a lottery can be thought of by the poor Tubers as a "good investment".

As soon as enough Actors are that dumbed-down, Public Officials can introduce even more obscene forms of taxation and regulation without causing riot. They can, for instance, force Workers to drive their cars to State Inspection Stations to be sure that their exhaust fumes aren't causing a totally unrelated and completely natural phenomenon that Media Frauds & Fraudettes have been told to label the "Ozone Hole" and speak about in worrisome tones.

Rather than fighting against such lunacy, brain-deadened Actors will fall back on the

"Hope Instinct" to believe that "This inspection may not be so bad, after all."

Then, More Enslavement

After the Public Officials have made the Workers and Civil servants drive to and pay for needless inspections every year, or quarter, or month, they will not stop. Not satisfied with even this amount of regulations destroying freedom, Public Officials will soon attempt to remove the right of all but the very rich or powerful to even have cars. When that happens, the distorted Hope Instinct will continue to keep us full of hope. We'll believe that we'll somehow be exempted.

Enslavement Brings Revolution

Periodically, oppressed Actors will attempt to be free. These attempts provide excuses. "We've got to crack down" on these vicious "radical elements" who are "endangering a free society".

When the full might of an armed state comes to bear on Freedom Seekers, their numbers dwindle quickly. Their leaders are removed from the Stage, and the rights of everyone else are trampled.

Leading to State Barracks and Dining Rooms

All who love the Guild of Government know that a most satisfying stage of development occurs when every Worker, Civil Servant, and Tuber is forced to live in State Barracks and eat in State Dining Rooms. Actors and their children are then completely available to satisfy all the desires of the depraved GOGsters who engineered the whole process.

Every single action by Bad Actors in the Guild of Government is designed to force all other Actors into State Barracks and State Dining Rooms. ". . .for their own good."

The Good Actors in the Guild of Government know what the worst are trying to do. There are many, usually at, or slightly above, the Civil Servant level, who are still attracted to the notion of truth and freedom. They will undermine, sabotage, expose, and short-circuit excessive GOG Power-grabs.

The smartest Civil Servants support Pro-Freedom action. They are a brave group of stalwarts. One reason that they secretly sub-vert the march into State Barracks and Dining Rooms is that they know who's going to get it next.

And, Finally, Death

The incredible Perversion of Hope keeps Workers from attacking Public Officials even after they've begun wholesale slaughter. "Well, Stalin is only killing ninety-nine out of every hundred of us Ukrainian farmers. There is a chance he won't get me." Ukrainian farm-ers used to say to themselves as they were herded into State Barracks and State Dining Rooms.

Up to the last moment of their lives, each be-lieved that the bullets in the guns of the firing squads would misfire, that they would fall alive into the trench, be overlooked and crawl away that night.

Hope springs eternal. The ongoing job of the GOG's Bad Actors is to twist that Hope into a mental disorder. It must be perverted to destroy the ability of the other Actors to make reasonable assumptions, form logical conclusions, and perform rational Acts.

As soon as Public Officials can bamboozle Actors into thinking they can profit by participating in Lotteries mathematically designed to impoverish them, they can more easily be enslaved. Then, murdered.

After the Workers

When they run out of Workers, Progressive Public Officials start killing Civil Servants. Then, Media Frauds & Fraudettes. Then, whatever Lubricators are left. Then, the rest.

The final scenes envisioned by the worst GOGster shows one last Public Official, him. That body, mind, and soul is bloated beyond human recognition, fiddling while Rome burns, the final flames fueled by the portly corpses of lesser GOGsters.

And, it all begins with making Actors believe that Lottery-Think is a rational type of thought.

Originally, of course,
Lotteries were illegal.

Lotteries used to be called "Numbers". They were run by L'il Grabbers who had jobs in what was called "Organized Crime". Recently, the lines between L'il Grabbers and Progressive Public Officials have become blurred. No one notices as Public Officials take over the operations of criminal activities and jail the criminals who used to run them.

Using their ever-obedient, ever-lying Media Frauds & Fraudettes, Public Officials now have more Civil servants and Workers than ever buying Lottery Tickets. How Public Officials love to watch Workers bankrupting themselves buying Lottery Tickets! Every dollar they waste makes them more dependent on Public Officials.

Daily on our Tubes, we can watch Public Officials spend vast amounts of money to have Media Frauds & Fraudettes tell Civil servants and Workers about how other Actors are winning vast amounts of money by playing the Lottery.

So,

One Attack on Life, Liberty, and the Pursuit of Happiness begins by replacing Truth and Logic with Wishful Thinking. This is done by making the Lottery Mentality seem reasonable and respectable.

In the endless battle for intellectual supremacy between objectivity and subjectivity, Lottery-Think makes subjectivity the clear winner.

And, don't the most Progressive kind of Public Officials love that?

IX

The Bad Actors

The Worst of the Bad Actors are The Dirty Old Men

The Very Worst of the Bad Actors are, and were, the Dirty Old Men. Dirty Old Men are infatuated with the things On-stage. They like(d) to pile up lots of Props. And, power. They put growing piles of Props in ever-vaster houses. They liked to hire Young Girls to work for them cheaply.

Some Dirty Old Men like to have sex with Young Girls. At first, it was hard for them to find Young Girls. The better kind of Young Girls were attracted to Young Men closer to their own age. They knew that the best thing for them to do was to marry Young Men and have families. The better class of Young Girls wouldn't waste themselves having sex with Dirty Old Men.

The Invention of War

The Dirty Old Men made an amazing discovery: They could get the most Decent Young Men to go out and kill each other for "idealistic reasons". That reduced competition for Props and Young Girls.

Killing off lots of Decent Young Men left Young Girls with no one to marry. When enough Dirty Old Men got worried about too many Decent Young Men competing with them, they would send word to each other:

"Let's kill a bunch of Decent Young Men so we can get more Young Girls and keep a firm grip on all our Props."

"This means war! Hooray!"

Innumerable Young Men would be killed. Afterwards, Dirty Old Men could afford bigger houses to hold more Props and inexpensive maids and nannies. Other services could be provided by the innumerable Young Girls who would never have families.

Then, World Wars

When Dirty Old Men get rid of Decent Young Men on one or two sections of The Stage, it is called a War. When they do it on several sections of The Stage at once, it is called a World War. In what they call World War I, the Dirty Old Men got very efficient. They used machine guns. And, poison gas.

Dirty Old Men who ran the sections called "England" and "Germany" loved to line up millions of Decent Young Men, and march them right into each other's machine guns. In the years following that War, the Dirty Old Men got richer than ever before. And, had lots of inexpensive maids, nannies, and secretaries. More Young Girls were made available than ever before.

The Dirty Young Men

In each performance, some Dirty Young Men figure out what the Dirty Old Men are doing. "I want to have lots of Props and Young Girls when I get old!" Dirty Young Men say to themselves. When a War came, they would avoid serving in the military so they could get old enough to become Dirty Old Men.

Dirty Young Men become the Dirty Old Men who cause the next war. Dirty Men gravitate toward becoming, or controlling, Public Officials.

Dirty Men Discoveries

Dirty Men discovered something that helped them: When Decent Young Men were busy killing each other, all the other Actors On Stage were more easily distracted by lies. Their enemies could be more easily removed from The Stage.

And, they could make money. *"We (you) have to fight this important War. We (you) need good weapons. We will make them and sell them to you. We will sell you food and clothes and lots of other things that We (you) really need to have."*

Dirty Men would have their Media Frauds & Fraudettes wave flags, sing songs, and sell bonds as they Exited more Decent Young Men and accumulated more Young Girls and Props.

And, raised taxes.

The Dirty Old Mens' #1 Desire: More Props, Fewer Actors

Dirty Old Men don't want other Actors to have children. "Parents will give money to their children instead of to us. We must give them contraceptives so that they will spend more time having sex but have fewer children.

"They should get rid of their babies before they are born. We must convince them that Overpopulation Is a Problem."

Actors know that the Dirty Old Men hate them and their children. "I'm having lots of kids!", say the ones who are able to keep from getting addicted to a few paltry Props. The Invisible Producer always helps them.

The Dirty Mens' #2 Desire: Make Other Actors Dumb

"Schools should only teach young Underlings to have sex and do work for us. They must not think. They must not read or write. They must not know how to compute. We must run the Public Schools to turn out only incompetents. "And, half of them will be Young Girls."

The Dirty Mens' #3 Desire:
Destroy the Idea of Right and Wrong.

"Do not teach Young Underlings about good and bad, right and wrong, truth or lies. Tell them that 'everything is relative'. Send them to the Public Schools.

"Hire men who love lies and fraud. Call them 'Doctors' so that everyone will know. Pay them more than they can make anywhere else. They will dumb down the children or be fired and have to get real jobs where they might get their hands dirty. We will hire Doctor Stooges who'd rather destroy children than that."

The Dirty Mens' #4 Desire:
That Actors do not Think
About God, Family or Country

Dirty Men do not want the Actors to think of themselves as part of something that could claim loyalty and taxes that the Dirty Men want for themselves. "They must think of themselves as part of One World. That makes it hard for them to see that we want to enslave them and destroy them and use them for whatever we want."

More Wars, More Props, Forever

When Dirty Old Men get their plans in place, their Productions can go on for centuries. The Actors rarely realize what is being done to them after they are dumbed down by the Dirty Men's Schools. There, the children can only learn that their sole hope of success is to Win The Lottery.

In an Earlier Performance, Dirty Old Men in Italy named their Production "Rome". It was a huge, well-organized Production. It provided Dirty Old Men with Props, wars, and Young Girls for centuries. It replaced several older, smaller Productions in Greece, some of which differed slightly in that they produced Young Boys for Dirty Old Men.

Two Kinds of History:

History is written two ways. Good Guy History tells stories of Actors protecting families and props from the Dirty Old Men who lust for both.

Good Guy History has heroes. Beowulf. Moses. Hercules. David. Robin Hood. George Washington.

Bad Guy History makes heroes of people who destroy families and freedom. Big Heroes in Bad Guy History are Pharaohs, Alexanders, Caesars, and Napoleons. Their little heroes are Bill Clinton and Al Gore.

Part X

How Many Sections
of the Stage are There?

There are three main sections of The Stage. One is run by White Actors. One is operated by Yellow Actors. Another is controlled by Black Actors. Each Group descends from one of the three sons of Noah, a prominent Mover/Storer in an Early Performance.

The three main sections of The Stage are divided into Sub-Stages. Some of them are further divided into smaller Sub-Sub Stages, and these are further subdivided. The smallest Sections are called "Villages" or "Boroughs" or "Wards" or "Collectives".

There are lots of Sections on the Stage because lots of Actors want, and need to have the opportunity, to be Public Officials. The Invisible Producer wants to give lots of Players the opportunity to exercise Free Will when it comes to lying to, stealing from, and killing their neighbors.

Why are Actors Packaged
in Different Colors?

Different packaging makes it easy to tell Actors apart. And, different Types and Colors indicate some, but not all, of their other attributes. Style and Color similarities encourage Actors to join into small groups and differentiate themselves from other groups.

The Invisible Producer made so many Styles and Colors to, First: Let Actors have the fun of being in communities of similar Actors. Second: To provide All Actors with opportunities to overcome deep dislikes and ancient hatreds for members of other communities.

The Stage is set up so that at any given time, any Actor or Group of Actors can choose to think that they have good reasons to feel animosity toward other Models and Colors of Actors.

Actors in Darker Packaging usually have fewer Props than Actors in Pale Packages. Some can be manipulated to resent that. Actors with a greater amount of Props can be made to resent the efforts continually being made to deprive them of their Props.

Many opportunities for dislikes exist.

The Son of the Invisible Producer told Actors that they would overcome these dislikes, if they knew what was good for them. Since the Three Invisible wanted to give a real test of each Actor's ability to overcome dislikes during billions of "My Life" playlets, He lets some of the Models have tendencies to do very dislikeable things.

Actors are produced in different Models and Colors to see if actors are able to overcome their dislike for Actors in different packaging before Final Review.

Do the Different Models and Colors of Actors have Different Abilities?

Yes. If abilities were all the same, there would be less meaningful amounts of dislike to overcome. Each and every Model of Actor has been given different mental, physical, emotional, and spiritual abilities. Each individual Actor has different Gifts.

Each individual Actor has what seem to be good reasons to dislike some Actors with gifts that may conflict with his own. Those dislikes, too, must be overcome.

What are the Sections of The Stage Today? Where are We?

Most Actors live on either the Black, White, or Yellow Section. Some Sections between and within those Sections are Mixed. North Africa is a Mixed Section. It is Tan.

The richest Sections of the Stage, where the most Actors have the most Props, are places where the Actors are the freest. As different Sections of the Stage obtain more Freedom, Actors who perform there get more Props. To create Problems to justify the existence of their Supporting Casts, Public Officials encourage poor Actors to move to richer Areas. Actors tend to move where they can get the most Props.

Many Actors want to move to the White Section. That's where the most Freedom is, so that's where the most Props are.

They Don't Want Us to Know That We are Only Actors on A Stage

Public Officials do not like to simplify The Stage. Once Actors understand that they are individual souls (ether discs) wrapped in different packaging, it is harder to get them to hate each other. When Actors see that possessions are only Stage Props, they may not take them seriously enough to be made to fight over.

If Actors cannot be made to steal and kill, the Dirty Men have a harder time getting them to remove each other from The Stage.

Once an Actor says: *"I was made in the image of God, and so were all the other Actors."* he is far less likely to hate the other Actors.

Things Are Made to Seem Complicated

Dirty Old Men don't want Actors to know that they are simply Actors. They don't want Actors to know that the Invisible Producer made the different Models and Colors for reasons of His own.

For centuries, Dirty Old Men have subsidized a view of the world so bizarre that it can barely be described with a straight face.

They have convinced many of the Actors that The Stage is fifteen or twenty billion years old. They have forced the Teachers in their schools to teach the children that they are little animals, whose ancestors were monkeys.

They have a funding structure that provides Props and power to people who say that they believe their silly theories.

Bad Actors definitely do not approve of the notion that the world is just a Stage made out of 3-D Fractals in a week by someone that much smarter than they are.

And, they don't like simplifying Sections of Stage so that everyone can see how obvious everything is.

The Yellow Section of the Stage

The Yellow Section has rarely been a good place for Actors who want to write and direct their own performances. In each performance, the most brutal and power-crazed of the Yellow Dirty Men have achieved total control of almost the entire Yellow Section. In each performance, Yellow Dirty Old Men force millions of Actors to make Involuntary Exits from The Stage. Now, they are beginning to Exit each other.

The Yellow Section is Very Depressing

The Yellow Dirty Men have killed or enslaved billions of their Actors. Their slaves are forced to make Doodads in bare sets with no props. They sell slave-made Doodads to people who don't mind.

White Dirty Men buy Doodads cheap in the Yellow Section to keep from having to pay decent wages to their Workers in the White Section. White Dirty Old Men want their workers to be Slaves, too. Yellow Dirty Old Men are glad to help.

There are <u>few</u> <u>Tubers</u> in the Yellow Section, and not many Media Frauds & Fraudettes. Sometimes, those in the White Section wish that the Yellow Dirty Old Men would take over the White Section for a while. "Get rid of a lot of useless Grabbers." they say to themselves. When they think a little longer, they understand that the Yellow Dirty Men will force everyone with an IQ over 110 to exit The Stage.

The smartest Yellow Actors have been hunted down and exterminated for so long that obedience and rote memory have largely replaced iniative and imagination on the Yellow Sections.

Those among them with High Brain Ranks either become Public Officials or are killed by them. Or, move to a White Section.

The Black Section of The Stage

The Black Section of The Stage has never let Actors excel at Doodad production. Actors on the Black Section spend most of their time trying to keep from being Exited, trying to Exit the Actors around them, trying to get into the White Sections, or getting Props and Death Doodads from White and Yellow Dirty Men who like to see Actors on the Black Section exit each other. "It's good for the environment!" they chuckle to each other as Black Actors turn their new machine guns on each other "More room for wildlife."

Black Actors who live in White Sectors of The Stage deny that Actors in the Black Section still own Slaves. They also deny the awfulness of conditions in the Black Section. They cannot bear to think of why the Black Section hasn't done well on its own.

Until White and Yellow Actors made entrances onto the Black Section, Black Actors mostly had stone, wood, and leather Doodads.

Tan Actors from Mixed Zones were able to buy and enslave whole Troupes of Black Actors from Black Public Officials for handfuls of shiny glass and metal Doodads.

Today, Blacks who live in White Sections sell millions and millions of their own votes to the worst kind of Public Official. In return, they still get Freebies that continue to enslave themselves, their children, and their relatives.

The White Section of The Stage

Actors in the White Section are freest to write their own lines and to direct their own performances. White Actors have removed Public Officials from The Stage just for trying to take away their freedom.

White Dirty Old Men, like the other Dirty Old Men, hate the idea of freedom. They work hard to enslave other Actors. They have found it's more profitable to enslave with taxes and regulations than with whips and chains.

Actors in the White Section are losing freedom rapidly. Public Officials work to channel income flows out of the Private Sector and into the Public Sector. More and more Actors on the White Section are being put under the thumbs of the White Dirty Men.

The Guild of Government grows, while the Free Market that supports it is being Undermined, and the day of reckoning draws near. Glad about it, White Dirty Men are diligently working to make things worse by destroying families and freedoms.

Actors in the Freest Section
Follow Different Directions

The most productive area on The Stage is the freest. America is wealthier than the other Sections of The Stage because Americans still have more freedom than Actors on any other section of The Stage.

For hundreds of years, White Actors on the American Section of The Stage produced Props and provided Performances according to rules and regulations which needed surprisingly little attention from Public Officials. Only recently has the GOG been able to undermine freedom in America by replacing the old ways with new, self-serving, petty, micro-managing that purposely undermines the importance of each individual Actor.

Old Rules for Private-Sector Workers in the USA Section:	GOG's New Rules for the USA Section:
1. Tell the truth.	1. There is no truth.
2. Work hard.	2. Do less.
3. Do what you say you will do.	3. If it feels good. . .
4. Don't hurt your neighbor.	4. Look out for #1.
5. The customer is always right.	5. Make them obey.
6. Be responsible and self-reliant.	6. Depend on us.

Meddlers hate freedom. Meddlers hate the Americans' Constitution because it defends freedom.

Why do Bad Actors Hate Freedom?

The White Sections of The Stage are the most complicated because Actors there are free to do more. The White Sections produce the widest variety of Props. Actors on the White Section Perform in intricate, specialized Roles that don't exist on the vastly simpler Black and Yellow Sections of the Stage.

Frankly, some people just aren't willing to put forth the effort to understand the complexities of The White Section. Actors whose minds are crippled by the Government Schools are easily convinced that they have a right, if not a duty, to get Props from others.

Most Freedom-Haters were made, not born, dumb. They were cleverly convinced that Free Lunches are respectable, and they all want one.

Why There are Bad Actors?

The Invisible Producer wants to see how well each Actor can overcome a desire for hatred and revenge.   He has told us very plainly that Revenge is His job. When we try to get revenge, we violate the Commandment He put First. To pass the Final Review, the first step we must take is to forgive.   Each of us can be grateful for having been given so many opportunities to take lots of such steps.

When the Invisible Producer's Son visited The Stage, He made one thing very clear: We must even love those who have hurt us very badly.

Loving hurtful people is a bizarre act.   It makes some of them think about the Invisible Producer and where they may be directed when they Exit. Ultimately and finally, people who hurt us are there so that we can love them.

We are allowed to hate things they choose to do. We can despise the twisted, self-serving ways they choose to think. We can deplore the hate that drives them to destroy life and freedom.

But, the only thing that can help them is our love. It is sad for them that they have chosen to lie and cheat and steal and kill. Loving them helps us get past The Critical Review.

No Justice Without Forgiveness

If we do not love and forgive those who hurt us, at the Final Review they will most certainly whine to The Critic: "I sinned because no one loved and forgave me."

If that's enough of an excuse, our lack of love and our inability to forgive will be what allows them to get off Scot-free.

So?

It's easiest to see what we see as a Stage peo-
pled with Actors and decorated with Props.
Simplification eliminates whole libraries full
of confusing distractions.

The Stage and its seeming complexities are
only here to give Actors the opportunities to
make the choices they need to save their souls.
The more clearly we see it, the simpler it is to
choose what's right.

There is no good reason for envy, anger, un-
kindness, and all that flows from such things.

Not unless an Actor is overly fond of Props.

Portersville, Pennsylvania.
1997

About the author:

Bill spent a critical year at St. Bede's in Peru, IL. There, he discovered, and never forgot, Chesterton.

After graduating from Ripon College and the Infantry Officers' Candidate School at Fort Benning, he served in the US and in South Vietnam.

He got an advanced degree in Library and Information Science from the University of Pittsburgh, and worked as a Children's Librarian.

As a Children's Librarian, he re-learned the Fairy Tales. And, started to believe them. After inheriting ten thousand dollars from his Grandfather, he left his tenured position and started a business.

Today, Adams Mfg. produces hundreds of household items and consumer products.

Bill has been re-elected to his School Board, which now has the lowest taxes in his County. "Cutting taxes is unbelievably difficult." he discovered. Duhh.

"All the World's a Stage" is his fourth book.

Bill Adams books from Old Drum Publishing:

Crats!
330 pages $6.95

A Pilgrim's Progress for our times. Modern pilgrimages through the murderous lunacy of Government programs to the peaceful sanity of The Church. Introduces the concepts that our Ancestors began in the Twelve Tribes. Points out that Conventional Reality is only that.

Combines Free Will and Fractals in a novel form. Gives credibility to a Creation only five or ten thousand years old. Shows how fossils could have been made in place to provide free intellectual choice between believing in God or in accidents.

Guarantee: When you've finished, you'll be able to tell people how real, live camels can go through the eyes of needles. What many think of as "bizarre" parts of the Bible will seem reasonable. "Prophetic".

My Yoke is Easy, and My Burden is Light . . .
112 pages. Illustrated. $6.95

Even people who've studied Scripture for years don't know the meaning of "my burden is light". They won't know unless you tell them. Introduces the exciting new concept of Current Technology Translation, which shows that Science is the servant of Faith. Puts computer-literate beginners ahead of many "experts" when it comes to bringing Moderns closer to God. Wonderful new ways to consider the origin and structure of matter.

All the World is a Stage
140 pages. Illustrated. $6.95

Simplifies the needless complexity of modern political/business/scientific/social theory. The world is, or can be reasonably seen as, a Stage. The only things on The Stage are Props and Actors. Props may have been made out of 3D Fractals programmed by God or His angels. There are Basic Props and Doodads.

Actors are beings that appear to have Free Will. Actors are in Troupes. Troupes are in Guilds. Whole libraries of theory are boiled down to let readers see far more clearly how they fit into reality and the Conventional Reality imposed on actors by Public Officials.

People ask: "Where can we find these Old Drum books? We know people who'd enjoy them."

And: "Why aren't you ever on talk shows with these new, exciting ideas?"

When you consider this book's comments about Media Frauds & Fraudettes, the usual TV appearances by the author are unlikely.

The best way to introduce new ideas is word of mouth. One by one, two by two, that's how the world is changed. You may know people whom you'd like to introduce to this new, simple approach to the Faith of our Fathers. *Books by the Dozen* makes it easy, and economical.

Books by the Dozen offers quantity discounts. A dozen or more of our $6.95 books are just $3.95 each. Simply mix or match any selections that total over a dozen books and add $3.50 p&h onto the total.

Need six to twelve copies? $4.95 each. $2.50 p&h to the total.

We also encourage you to ask your favorite bookstore to make these books available.

Phone Orders: 1-800 OLD-DRUM with Visa, MC.
Fax: (412) 368-3339 Other calls: (412) 368-3338
Postal Orders: Old Drum Publishing, Box 401,
Portersville, PA   16051-0401

NAME_____

ADDRESS_____

CITY_____STATE____ZIP_____

Telephone_____ Fax_____

Please send the following at $6.95 ea.   (1-6 copies):

> _____copies of *Crats!*
>
> _____copies of *My Yoke is Easy*
>
> _____copies of *All the World is a Stage*
>
> _____ total copies at $6.95 ea = _____

Add $2.00 p&h for 1-6 books. $2.00

PA residents, please add 6%. Order Total:_____

Buying presents? Use quantity discounts to get something good for people you like. *Books by the Dozen:* 12 or more books, mix or match, $3.95 per book + $3.00 p&h. .

> ____ copies of *Crats!*
>
> ____copies of *My Yoke is Easy*
>
> ____copies of *All the World is a Stage*
>
> _____ Total copies at $3.95 ea = _____

Add $3.50 p&h for 12 or more books. $3.50

PA residents, please add 6%. Order total:_____

6-12 books, $4.95 per book plus $2.50 p&h.. PA. residents, please add 6%.

All The World's A Stage . . .

Phone Orders: 1-800 OLD-DRUM with Visa, MC.
Fax: (412) 368-3339 Other calls: (412) 368-3338
Postal Orders: Old Drum Publishing, Box 401,
Portersville, PA 16051-0401

NAME_____

ADDRESS_____

CITY_____STATE____ZIP_____

Telephone_____Fax_____

Please send the following at $6.95 ea. (1-6 copies):

    _____copies of *Crats!*

    _____copies of *My Yoke is Easy*

    _____copies of *All the World is a Stage*

    _____ total copies at $6.95 ea= _____

Add $2.00 p&h for 1-6 books. $2.00

PA residents, please add 6%.Order Total:_____

Buying presents? Use quantity discounts to
get something good for people you like. *Books by the
Dozen:* 12 or more books, mix or match, $3.95 per
book + $3.00 p&h. .

    _____ copies of *Crats!*

    _____copies of *My Yoke is Easy*

    _____copies of *All the World is a Stage*

    _____ Total copies at $3.95 ea= _____

Add $3.50 p&h for 12 or more books. $3.50

PA residents, please add 6%. Order total:_____

6-12 books, $4.95 per book plus $2.50 p&h.. PA. residents, please add 6%.

All The World's A Stage . . .

. . . & Every Actor Meets The Critic.

All The World's A Stage . . .